T0383093

PPG Signal Analysis

PPG Signal Analysis

An Introduction Using MATLAB

Mohamed Elgendi

CRC Press
Taylor & Francis Group
Boca Raton London New York

CRC Press is an imprint of the
Taylor & Francis Group, an **informa** business

MATLAB® is a trademark of The MathWorks, Inc. and is used with permission. The MathWorks does not warrant the accuracy of the text or exercises in this book. This book's use or discussion of MATLAB® software or related products does not constitute endorsement or sponsorship by The MathWorks of a particular pedagogical approach or particular use of the MATLAB® software.

First edition published 2021
by CRC Press
6000 Broken Sound Parkway NW, Suite 300, Boca Raton, FL 33487-2742

and by CRC Press
2 Park Square, Milton Park, Abingdon, Oxon, OX14 4RN

© 2021 Taylor & Francis Group, LLC

CRC Press is an imprint of Taylor & Francis Group, LLC

Reasonable efforts have been made to publish reliable data and information, but the author and publisher cannot assume responsibility for the validity of all materials or the consequences of their use. The authors and publishers have attempted to trace the copyright holders of all material reproduced in this publication and apologize to copyright holders if permission to publish in this form has not been obtained. If any copyright material has not been acknowledged please write and let us know so we may rectify in any future reprint.

Except as permitted under U.S. Copyright Law, no part of this book may be reprinted, reproduced, transmitted, or utilized in any form by any electronic, mechanical, or other means, now known or hereafter invented, including photocopying, microfilming, and recording, or in any information storage or retrieval system, without written permission from the publishers.

For permission to photocopy or use material electronically from this work, access www.copyright.com or contact the Copyright Clearance Center, Inc. (CCC), 222 Rosewood Drive, Danvers, MA 01923, 978-750-8400. For works that are not available on CCC please contact mpkbookspermissions@tandf.co.uk

Trademark notice: Product or corporate names may be trademarks or registered trademarks, and are used only for identification and explanation without intent to infringe.

Library of Congress Cataloging-in-Publication Data
Names: Elgendi, Mohamed, author.
Title: PPG signal analysis : an introduction using MATLAB / Mohamed Elgendi.
Other titles: Photoplethysmogram signal analysis
Description: Boca Raton : Taylor & Francis, [2018]
Identifiers: LCCN 2018018138 (print) | LCCN 2018018683 (ebook) | ISBN 9780429449581 (ebook General) | ISBN 9780429831126 (Pdf) | ISBN 9780429831119 (ePUB) | ISBN 9780429831102 (Mobipocket) | ISBN 9781138049710 (hardback : alk. paper)
Subjects: | MESH: MATLAB. | Photoplethysmography | Signal Processing, Computer-Assisted | Image Interpretation, Computer-Assisted--methods | Algorithms | Models, Theoretical
Classification: LCC R857.O6 (ebook) | LCC R857.O6 (print) | NLM WG 141.5.P7 | DDC 616.07/54--dc23
LC record available at https://lccn.loc.gov/2018018138

ISBN: 978-1-138-04971-0 (hbk)
ISBN: 978-0-429-44958-1 (ebk)

Typeset in Minion
by SPi Global, India

To Halla, Maya and Izac.

Contents

Figures and Tables

Preface

OVER THE LAST FEW YEARS, THE PHOTOPLETHYSMOGRAM (PPG) SIGNAL HAS attracted many researchers to investigate cardiovascular diseases, given that it contains valuable biological information about the body. Currently, no book exclusively discusses the theory and practice of PPG signals in medical applications, and this book is an attempt to address these applications in a simple and easy to understand manner.

Several PPG examples discussed in this book were also used while I was delivering the 'Biosignals and Systems' course at the University of British Columbia. During the delivery of the course, hands-on lab experience was provided to students on biosignals using MATLAB software. Because of the positive feedback received, the idea of creating a book specifically discussing PPG signals began.

Different labs and universities have varying preferences in terms of using software for analysis. Some prefer freely available software such as Python, rather than commercial software such as MATLAB. However, the reason MATLAB was used here was because of its substantial number of functions and apps. More importantly, it is easier for beginners to understand the concept using MATLAB, as it includes all the toolboxes needed. At the same time, other software requires the installation of extra packages and an IDE. MATLAB also has a broad scientific community; it is used in many universities (although few companies have the money to buy a license).

As an educator, teaching students from different backgrounds is not an easy task. Therefore, I believe that the use of MATLAB is more effective, as it enables students to progress quickly during the problem-solving process. Students are focused on using MATLAB's toolboxes instead of wasting time trying different packages and IDEs. My experience with MATLAB for teaching students from different backgrounds is satisfying and encouraging.

The book comprises 11 chapters, each with several sections. The first section outlines the learning objectives of the chapter. Then, each subsequent section discusses the fundamental concepts of the topic, followed by MATLAB examples. Readers who are familiar with MATLAB may skip some MATLAB commands; however, the ability to see and visualize the PPG examples will inspire their current and future PPG applications. Note that Chapter 11 provides a number of publicly available databases that can be downloaded and used freely for testing and evaluation.

To maximize the benefits of this book, you will need to try all examples, modify them, and check how they can be used in your application. Underestimating a simple concept or MATLAB example may not be a good idea. Concepts and examples were simplified to suit different readers from different backgrounds so that engineers, computer scientists, clinicians, and psychologists can start from scratch and reach a point where all are on common ground. I hope that I succeeded in achieving this. If not, then please consider it as an attempt.

I could not have written this book without the continuous support of the Department of Obstetrics and Gynecology and the Department of Electrical and Computer Engineering at the University of British Columbia, and the BC Children's Hospital Research Institute. First and foremost, I would like to thank my family for their kind support and endless patience, as well as their continued encouragement. Finally, I would like to thank you, the reader, for your interest in this topic, and I sincerely hope that you find this material valuable towards developing your application, whether it is a mobile app or a wearable device. One request, your feedback on this book would be greatly appreciated.

Acknowledgments

THE AUTHOR IS GRATEFUL FOR THE SUPPORT OF MINING FOR Miracles, the BC Children's Hospital Foundation, and the Women's Health Research Centre of British Columbia, Vancouver, British Columbia, Canada.

I would like to thank the reviewers John Allen (Newcastle University, UK), Dingchang Zheng (Coventry University, UK), and Richard Fletcher (Massachusetts Institute of Technology, USA), for their valuable feedback, which led to improved content. Also, I would like to thank my PhD students Yongbo Liang and Tang Qunfeng, for checking the clarity of the book and for providing valuable suggestions.

The Author

Dr. Mohamed Elgendi is currently a Senior Postdoctoral Fellow at UBC's Department of Obstetrics and Gynecology, an Adjunct Professor at UBC's Department of Electrical and Computer Engineering, a Senior Member at IEEE, and a Senior Fellow at the Brain Sciences Foundation. In addition to his more than ten years of experience in the field of data analysis, he received training on Big Data Analysis and Leadership in Education from MIT.

Dr. Elgendi's experience in the areas of biomedical signal processing, digital health, data analysis, and visualization includes his work in Global Health with the PRE-EMPT Initiative (funded by the Bill and Melinda Gates Foundation), the Institute for Media Innovation at Nanyang Technological University (Singapore), and Alberta's Stollery Children's Hospital (Canada). Dr. Elgendi specializes in bridging the areas of engineering, computer science, psychology, and medicine for knowledge translation.

How to Use This Book?

E ACH CHAPTER TYPICALLY STARTS WITH AN OVERVIEW OF THE TOPIC. A broad perspective is usually accompanied by mathematical equations to support the concept. Some chapters are more theoretical than others, depending on the nature of the topic. For example, Chapter 7, on Event Detection, provides the theoretical methodology for detecting events in PPG signals, while Chapter 10, on Global Health, discusses a six-step framework for PPG applications. At their core, most chapters include MATLAB codes, and examples are discussed.

All MATLAB functions and codes are created and tested on the latest version of MATLAB (currently R2018b). Reference is made to obsolete MATLAB functions; however, most of the functions used in this book are compatible with older versions of MATLAB.

CHAPTER 1

Math Foundations

This chapter gives an overview of matlab and its basic arithmetic operations and functions, as well as a short introduction to matrices and matrix manipulation. Note that the MATLAB codes and examples are presented in a different font, and each line is numbered as follows:

```
1  >> x = [1 2 2 4 4 5 5];
2  >> figure;
3  >> plot(x);
```

1.1 LEARNING OBJECTIVES

The learning objectives of this chapter are to:

- Learn basic MATLAB functions, such as addition and subtraction
- Learn how to store numbers, strings, and logical values as variables
- Learn how variables can be imported and exported

1.2 SCALARS

1.2.1 Scalar Mathematical Operations

Scalar mathematical operations are applied on quantities that are fully described only by a magnitude (or a numerical value).

Addition ex1: Adding two positive numbers (5,2) is performed as follows:

```
1  >> 5 + 2
2
3  ans =
4        7
```

Note that the spacing between each number and operator does not matter. For example, multiple spaces between 5 and 2 in the addition process will still provide the correct answer:

```
1  >> 5        +        2
2
3  ans =
4        7
```

Addition ex2: Adding two negative numbers (−5,−2) is performed as follows:

```
1  >> (−5)  +  (−2)
2
3  ans =
4        −7
```

Note that if the parentheses are removed, the operation will still be calculated correctly:

```
1  >> −5 + −2
2
3  ans =
4        −7
```

Subtraction ex1: Subtracting a positive number (2) from a positive number (5) is performed as follows:

```
1  >> 5−2
2
3  ans =
4        3
```

Subtraction ex2: Subtracting a negative number (−2) from a negative number (−5) is performed as follows:

```
1  >> (−5)−(−2)
2
3  ans =
4        −3
```

Note that if the parentheses are removed, the operation will still be calculated correctly:

```
1  >> -5 - -2
2
3  ans =
4          -3
```

Multiplication ex1: Multiplying two positive numbers (5,2) is performed as follows:

```
1  >> 5*2
2
3  ans =
4          10
```

Multiplication ex2: Multiplying two negative numbers (−5,−2) is performed as follows:

```
1  >> (-5)*(-2)
2
3  ans =
4          10
```

Note that if the parentheses are removed, the operation will still be calculated correctly:

```
1  >> -5 * -2
2
3  ans =
4          10
```

Division ex1: Dividing a positive number (5) by a positive number (2) is performed as follows:

```
1  >> 5/2
2
3  ans =
4          2.5
```

Division ex2: Dividing a negative number (−5) by a positive number (−2) is performed as follows:

```
1  >> -5/-2
2
3  ans =
4          2.5
```

Power ex1: Raising a positive number (5) to a positive number (2) is performed as follows:

```
1 >> 5^2
2
3 ans =
4        25
```

Power ex2: Raising a negative number (−5) to a negative number (−2) is performed as follows:

```
1 >> (-5)^(-2)
2
3 ans =
4        0.0400
```

Natural Log ex1: Returning the log value of a positive number (5) is performed as follows:

```
1 >> log(5)
2
3 ans =
4        1.6094
```

Natural Log ex2: Returning the log value of a negative number (−5) is performed as follows:

```
1 >> log(-5)
2
3 ans =
4        1.6094 + 3.1416i
```

Exponential ex1: Returning the exponential value of a positive number (5) is performed as follows:

```
1 >> exp(5)
2
3 ans =
4        148.4132
```

Exponential ex2: Returning the exponential value of a negative number (−5) is performed as follows:

```
1 >> exp(-5)
2
3 ans =
4        0.0067
```

Sin ex1: Returning the sine of a positive value (pi/2) is performed as follows:

```
1  >> sin(pi/2)
2
3  ans =
4       1
```

Sin ex2: Returning the sine of a negative value (–pi/2) is performed as follows:

```
1  >> sin(-pi/2)
2
3  ans =
4       -1
```

1.2.2 Assigning Scalar Values

To assign a scalar value to a variable, simply introduce the variable to the lefthand side of the equals sign, and the scalar value(s) to the righthand side of the equals sign:

```
1  >> x = 1;
2  >> y = 2;
3  >> z = x + y
4
5  z =
6
7       3
```

1.3 VECTORS
1.3.1 Vector Mathematical Operations

Vector mathematical operations are applied on quantities that are fully described by a magnitude and a direction.

VectorAdd ex1: Adding a scalar value of 5 to a vector of [1 3].

```
1  >> 5+[1  3]
2
3  ans =
4
5       6     8
```

VectorAdd ex2: Adding two separate vector values of [5 2] and [1 3] to each other.

```
1  >> [5 2]+[1 3]
2
3  ans =
4
5        6    5
```

Note that both vectors must be of the same size.

VectorSub ex1: Subtracting a vector of [1 3] from a scalar value of 5.

```
1  >> 5-[1 3]
2
3  ans =
4
5        4    2
```

Note that the scalar value of 5 has been treated as a vector of [5 5].

VectorSub ex2: Subtracting one vector of [1 3] from another vector of [5 2].

```
1  >> [5 2]-[1 3]
2
3  ans =
4
5        4   -1
```

Multiplication ex1: Multiplying one vector of [1 3] by a scalar value of 5.

```
1  >> 5*[1 3]
2
3  ans =
4
5        5    15
```

Multiplication ex2: Element-by-element multiplication of one vector by another vector.

```
1  >> [5 2]*[1 3]
2  Error using *
3  Inner matrix dimensions must agree
```

Note that the element-by-element multiplication operator does not work as addition and subtraction. We have to add a dot (.) before the multiplication operator (*):

```
1  >>  [5 2].*[1 3]
2
3  ans =
4
5         5          6
```

Multiplication ex3: Dot product of two vectors of [5 2] and [1 3].

```
1  >>  dot([5 2],[1 3])
2
3  ans =
4
5        11
```

Division ex1: Dividing a vector of [1 3] by a scalar value of 5.

```
1  >>  [1 3]/5
2
3  ans =
4
5         0.2000          0.6000
```

Division ex2: Element-by-element division of one vector by another vector.

```
1  >>  [5    2]./[1    3]
2
3  ans =
4
5         5.0000          0.6667
```

Transpose ex: Linear algebraic transposition of an array.

```
1  >>  [5    2]'
2
3  ans =
4
5         5
6         2
```

Note that the apostrophe converted the row array into a column array.

To assign a vector to a variable, simply introduce the variable on the lefthand side of the equals sign, and the vector to the righthand side of the equals sign; this is performed as follows:

```
1  >> x = [1 2 3];
2  >> y = [2 4 9];
3  >> z = x + y
4
5  z =
6
7     3    6    12
```

1.3.3 Assigning Vector Elements Using a Function

There are three essential MATLAB functions that create a vector of a specific size and with defined spacing between the elements: *ones, zeros,* and *linspace.* These functions help with the initialization step of a variable. The *ones* and *zeros* functions have two inputs. One of these inputs has to be equal to 1, depending on whether the vector is going to be a row or a column. For example, here is how to create a row vector of length 4, filled with 1s:

```
1  >> x = ones(1 ,4)
2
3  x =
4
5     1    1    1    1
```

To create a column vector of length 4 that is filled with 0s, use:

```
1  >> x = zeros(4 ,1)
2
3  x =
4
5     0
6     0
7     0
8     0
```

The *linspace* function creates vectors with linearly spaced elements. Here is an example showing how to create a row vector that starts with 1 and ends with 4, and that is spaced to create five elements:

```
1  >> x = linspace(1,4,5)
2
3  x =
4
5      1.0000      1.7500      2.5000      3.2500
          4.0000
```

1.3.4 Assigning Vector Elements Using a Colon (:)

The use of a colon (:) is similar to the use of the *linspace* function discussed in section 1.3.3. The following is an example of how create a vector starting with 1, ending with 6, and stepping by two using a *colon*:

```
1  >> x = 1:2:6
2
3  x =
4
5      1    3    5
```

If we repeat the previous example with *linspace*, the output will be as follows:

```
1  >> x = linspace(1, 6, 2)
2
3  x =
4
5      1    6
```

There is a huge difference between the colon and *linspace*, even though they look identical. The *colon* operator creates a vector based on the specified increment/decremental step, whereas the *linspace* function creates a vector based on the required number of elements.

1.3.5 Addressing Vector Elements

To access the third element in a created vector, write the following:

```
1  >> x = linspace(1, 4, 5)
2
3  x =
4
5      1.0000      1.7500      2.5000      3.2500
          4.0000
6
7  >> x(3)
```

```
 8
 9  ans =
10
11      2.5000
```

We can use the *colon* to access the second, third, and fourth elements of a vector created by *linspace* as follows:

```
 1  >> x = linspace(1, 16, 8)
 2
 3  x =
 4
 5      1.0000    3.1429    5.2857    7.4286    9.5714
        11.7143   13.8571   16.0000
 6
 7  >> x(2:4)
 8
 9  ans =
10
11      3.1429    5.2857    7.4286
```

1.3.6 Increasing the Vector Size

If we need to add an element with a value of 10 to the vector, we can increase the vector size on the fly as follows:

```
 1  >> x = linspace(1,4,5)
 2
 3  x =
 4
 5      1.0000    1.7500    2.5000    3.2500    4.0000
 6
 7  >> x(6)=10
 8
 9  x =
10
11      1.0000    1.7500    2.5000    3.2500    4.0000
        10.0000
```

1.4 MATRICES

1.4.1 Matrix Mathematical Operations

Matrix mathematical operations are applied to quantities that are fully described by a magnitude and a direction.

Addition ex1: Adding a scalar value of 5 to a matrix of [3 4; 2 3].

```
1  >> 5+[3 4;2 3]
2
3  ans =
4
5        8     9
6        7     8
```

Addition ex2: Adding one matrix to another matrix.

```
1  >> [1   3;1   2]+[3   4;2   3]
2
3  ans =
4
5        4     7
6        3     5
```

Note that both matrices must be of the same size. Here is an example for adding a vector of [1 5] to a matrix of [3 4; 2 3]:

```
1  >> [1   5]+[3   4;2   3]
2  Error using     +
3  Matrix dimensions must agree.
```

An error has occurred because of a difference in size; the vector is 1×2, whereas the matrix is 2×2.

Subtraction ex1: Subtracting a matrix of [3 4; 2 3] from a scalar value of 5.

```
1  >> 5-[3   4;2   3]
2
3  ans =
4
5        2     1
6        3     2
```

Note that the scalar value of 5 has been treated as a matrix of [5 5; 5 5].

Subtraction ex2: Subtracting one matrix from another matrix.

```
1  >> [1   3;1   2]-[3 4;2   3]
2
3  ans =
4
5       -2    -1
6       -1    -1
```

Multiplication ex1: Multiplying a matrix of [1 3] by a scalar value of 5.

```
1 >> 5*[3   4;2   3]
2
3 ans =
4
5       15       20
6       10       15
```

Multiplication ex2: Cross-product of two matrices.

```
1 >> [1   3;1   2]*[3   4;2   3]
2
3 ans =
4
5       9       13
6       7       10
```

Multiplication ex3: Dot product of two matrices.

```
1 >> [1   3;1   2].*[3   4;2   3]
2
3 ans =
4
5       3       12
6       2        6
```

Note that the element-by-element multiplication operator does not work in the same way as addition and subtraction. We have to add a dot (.) before the multiplication operator (*). We can use the **dot()** function if the size of the matrices greater than or equal to 3.

Division ex1: Dividing a matrix of [1 3; 1 2] by a scalar value of 5.

```
1 >> [1   3;1   2]/5
2
3 ans =
4
5       0.2000       0.6000
6       0.2000       0.4000
```

Division ex2: Element-by-element division of one matrix by another matrix.

```
1 >> [1   3;1   2]./[3   4;2   3]
2
```

```
3 ans =
4
5        0.3333        0.7500
6        0.5000        0.6667
```

Transpose ex: Here is an example of the linear algebraic transposition of a matrix:

```
1 >> [1   3;1   2]'
2
3 ans =
4
5        1         1
6        3         2
```

Note the convertsion of the row arrays to column arrays.

1.4.2 Assigning Matrix Elements

To assign a matrix to a variable, simply introduce the variable on the lefthand hand side of an equals sign, and the matrix to the righthand side of the equation:

```
1 >> x=[1   3;1   2];
2 >> y=[3   4;2   3];
3 >> x+y
4
5 ans =
6
7        4         7
8        3         5
```

1.4.3 Assigning Matrix Elements Using a Function

We can use the two essential MATLAB functions *ones* and *zeros* to create matrices of specific sizes. Let us create a 4×4 matrix filled with 1s:

```
1 >> x = ones(4,  4)
2
3 x =
4
5        1        1        1        1
6        1        1        1        1
7        1        1        1        1
8        1        1        1        1
```

To create a 3 × 3 matrix filled with 0s use:

```
1  >> x = zeros(3, 3)
2
3  x =
4
5        0       0       0
6        0       0       0
7        0       0       0
```

Note that *linspace* only generates vectors; it does not generate matrices.

1.4.4 Addressing Matrix Elements

Here is an example showing how to access an element in the second row and third column in a created matrix.

```
1   A =
2
3         1       2       3
4         4       5       6
5         7       8       9
6
7   >> A(2,3)
8
9   ans =
10
11        6
```

We can use the *colon* to access all elements in the the second row of the created matrix as follows:

```
1  >> A(2,:)
2
3  ans =
4
5        4       5       6
```

1.5 RELATIONAL OPERATORS

Greater-than ex1: Returning the results of the comparison of a scalar value of (5) to a scalar value of (2) to investigate which is the greater.

```
1  >> 5>2
2
3  ans =
4        1
```

Note that the result of the comparison is 1, which indicates it is true that 5 is greater than 2.

Greater-than ex2: Returning the results of the comparison of a scalar value of (5) to a vector of [6 3 2 9 7].

```
1  >> 5>[6 3 2 9 7]
2
3  ans =
4
5      0    1    1    0    0
```

Note that the result is presented as a vector comprising binary numbers; 0 means "false", 1 means "true", after the comparison of each element in the vector with the scalar value of 5.

Greater-than ex3: Returning the results of a comparison of a vector to investigate whether it is greater than the vector [6 3 2 9 7].

```
1  >> [5 6 1 7 9]>[6 3 2 9 7]
2
3  ans =
4
5      0    1    0    0    1
```

Note that the comparison was carried out element-by-element from each vector in a vector. The binary numbers 0 and 1 represent the result of the logical comparison.

Operators such as less-than (<), greater-than-or-equal (>=), less-than-or-equal (<=), equal (==), and not-equal (=) can be used in a similar manner to the method shown in the previous two examples.

1.6 NaN

NaN is the IEEE arithmetic representation for *Not a Number* (NaN). It results from mathematical operations that have an undefined numerical value. Here are some examples that produce NaN:

```
1  >> −Inf+Inf
2
3  ans =
4
5      NaN
6
7
```

```
 8 >> 0*Inf
 9
10 ans =
11
12     NaN
13
14
15 >> Inf/Inf
16
17 ans =
18
19     NaN
20
21
22 >> 0/0
23
24 ans =
25
26     NaN
```

Keep the NaN operator in mind as, later on, we will see how NaN will play a major role in PPG signal analysis. MATLAB has a few built-in functions for working with NaNs; for instance, *isnan*—which returns a logical value of 1 in the event that the input is NaN. Here is an example:

```
1 >> isnan(NaN)
2
3 ans =
4
5       1
```

1.7 STRINGS

Strings are matrices with character elements. The normal rules of assignment and variable creation apply. When you assign a string to a variable, the string has to be enclosed in single quotes. The following example shows how to create string variables.

```
1 >> first = 'Derek';
2 >> last  = 'Abbott';
3 >> name  = [first,'  ',last]
4
5 name =
6
7 Derek Abbott
```

Note that the strings ('Derek' and 'Abbott') are vectors and therefore must be concatenated within square brackets. In order to have a space between the *first* and *last* variables, we need to put a space character between two closing quote marks (' '). The variable *first* is composed of five characters in a 4 × 1 vector, whereas the variable *last* is an 6 × 1 vector of characters. The third letter in variable *first* can be accessed as follows:

```
1  >> first(3)
2
3  ans =
4
5  r
```

MATLAB has many functions for working with strings. Here are a few examples:

- *int2str*: To convert an integer number into a string.

```
1  >> x=int2str(15)
2
3  x =
4
5  15
6
7  >> whos x
8    Name          Size              Bytes     Class
       Attributes
9
10   x             1x2                 4        char
```

Here, we see how the integer number is converted into a string. The variable *x* contains the string; however, it is difficult to distinguish between the number and the string. Therefore, we need to use the *whos* function to display the type of the variable. It is clear now that *x* contains a string as the class is "char". The size of *x* is 1 × 2, as the number contains two numbers and is now converted into two characters.

- *num2str*: To convert a real number into a string.

```
1  >> x=num2str(15.34)
2
3  x =
4
5  15.34
```

```
 6
 7  >> whos x
 8     Name                Size                    Bytes           Class
       Attributes
 9
10     x                   1x5                        10           char
```

- *lower*: To convert a string into a lowercase string.

```
 1  >> x='Nigel H. Lovell'
 2
 3  x =
 4
 5  Nigel H. Lovell
 6
 7  >> lower(x)
 8
 9  ans =
10
11  nigel h. lovell
```

- *upper*: To convert a string into an uppercase string.

```
 1  >> x='Gary Clifford'
 2
 3  x =
 4
 5  Gary Clifford
 6
 7  >> upper(x)
 8
 9  ans =
10
11  GARY CLIFFORD
```

- *strcmp*: To compare two strings and return 1 in the event that they are identical.

```
 1  >> x='UBC'
 2
 3  x =
 4
 5  UBC
```

```
 6
 7 >> y='ubc'
 8
 9 y =
10
11 ubc
12
13 >> strcmp(x,y)
14
15 ans =
16
17        0
```

- *strrep*: To replace all occurrences of a string with a different string.

```
1 >> strrep('Life', 'L', 'W')
2
3 ans =
4
5 Wife
```

- *findstr*: To return the starting indices of any occurrences of the shorter of the two strings.

```
1 >> findstr('Life', 'fe')
2
3 ans =
4
5        3
```

1.8 STRUCTURES

Scalars, vectors, and matrices are not the only types of data that MATLAB offers. Structures are also supported by MATLAB to create variable types that mix numbers, strings, and arrays. As an example, let us create a data structure that contains the information for a single participant. We will store the first name (string), last name (string), age (scalar), three blood pressure readings (vector), and the presence of hypertension is reflected in a logical value of 0 or 1 (logical) as follows:

```
1 >> Participant.FirstName='Kirk';
2 Participant.LastName='Shelley';
3 Participant.Age=45;
```

```
4  Participant.BloodPressure=[110, 140, 113];
5  Participant.Hypertension=logical([0,1,0]);
6  >> Participant
7
8  Participant =
9
10        FirstName:  'Kirk'
11         LastName:  'Shelley'
12              Age:  45
13   BloodPressure:  [110 140 113]
14    Hypertension:  [0 1 0]
```

To address a value within a structure, the name of the structure has to be followed, first, by a dot and then by the field name:

```
1  >> Participant.BloodPressure(2)
2
3  ans =
4
5     140
```

We can now add further participant information to the same structure:

```
1  >>    Participant(2).FirstName='John';
2  Participant(2).LastName='Allen';
3  Participant(2).Age=50;
4  Participant(2).BloodPressure=[139,  130,  141];
5  Participant(2).Hypertension=logical([1,0,1]);
6  >>    Participant(2)
7
8  ans =
9
10        FirstName:  'John'
11         LastName:  'Allen'
12              Age:  50
13   BloodPressure:  [139  130  141]
14    Hypertension:  [1  0  1]
```

Let us suppose we want to access the third blood pressure reading of 'John Allen':

```
1  >> Participant(2).BloodPressure(3)
2
3  ans =
4
5     141
```

1.9 CELL

A *cell* is a variable that is similar to *structure*, but it is more general and has a different notation. A *cell* can contain any variable from a scalar to a structure, or another cell. We can create a cell array, which is a matrix comprising cell elements. To claraify: the main difference between a cell and a structure can be summarized as follows:

A structure consists of a set of fields, whereas a cell is a matrix. This means accessing the cell elements (indexed by row and column numbers) is a more dynamic procedure than that required to access a structure (indexed by field names).

Figure 1.1 shows an example of a 3 × 3 *cell array*. It is clear that each cell contains a unique variable; for instance, the first row in the first column contains a matrix, whereas the third row in the third column contains a structure. Cells enable us to assign a variety of data types to one type of variable, thereby providing a powerful diagnostic tool.

There are two methods with which to create a cell array: cell indexing and content indexing.

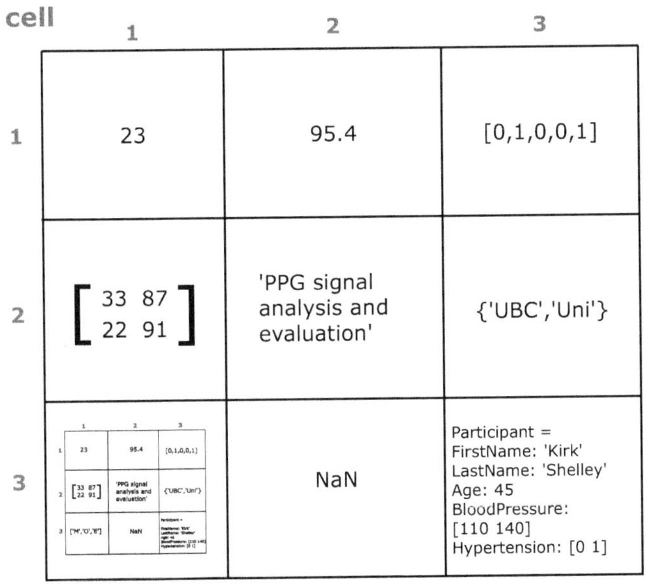

cell	1	2	3
1	23	95.4	[0,1,0,0,1]
2	$\begin{bmatrix} 33 & 87 \\ 22 & 91 \end{bmatrix}$	'PPG signal analysis and evaluation'	{'UBC','Uni'}
3	(nested cell array)	NaN	Participant = FirstName: 'Kirk' LastName: 'Shelley' Age: 45 BloodPressure: [110 140] Hypertension: [0 1]

FIGURE 1.1   An example of a cell array.

Cell indexing

```
 1  >> SubjectName={'Rafael Ortega','Andrew Reisner',
    'Gerald Dziekan'}
 2
 3  SubjectName =
 4
 5      'Rafael Ortega'          'Andrew Reisner'
        'Gerald Dziekan'
 6
 7  >> BloodPressure=[100    110    109],[140    145
    143],[100    101    100]
 8
 9  BloodPressure =
10
11      [1x3  double]          [1x3  double]      [1x3
        double]
12  >> Data={SubjectName,BloodPressure}
13
14  Data =
15
16      {1x3 cell}    {1x3 cell}
```

Content indexing

```
 1  >> Data{1,1}={'Rafael  Ortega','Andrew  Reisner',
    'Gerald Dziekan'}
 2
 3  Data =
 4
 5      {1x3 cell}
 6
 7  >> Data {1,2}={[100    110    109],[140    145
    143],[100    101    100]}
 8
 9  Data =
10
11      {1x3 cell}    {1x3 cell}
```

To access, for example, the first name in the data cell array—in both cases, cell and content indexings—content indexing will be used as follows:

```
1  >> Data{1,1}{1,1}
2
3  ans =
4
5  Rafael Ortega
```

1.10 IMPORT/EXPORT DATA

MATLAB has many built-in functions to load data variables from different file types, such as:

- Excel worksheet using *xlsread* function
- Common-separated numbers using *csvread* and *importdata* functions
- Text using *textsread* function
- Saved MATLAB workspace using *load* and *importdata* functions
- ASCII using *importdata* function
- Image workspace using *importdata* function
- Audio using *importdata* function

We can save variables from the workspace using MATLAB's built-in function *save* as follows:

```
1  >> a =[1,   2,   3,   4;23   24   25   26]
2
3  a =
4
5      1    2    3    4
6     23   24   25   26
7
8  >> b ='My first test'
9
```

```
10  b =
11
12  My first test
13
14  >> c = [0,1]
15
16  c =
17
18       0       1
19
20  >> save   test1
```

Note that the *save* function saved all variables in the workspace. If we needed to save one variable, such as *a*, then we write the save command line as:

```
1  >> save   test2   a
```

As you can see, we saved two files: test1 and test2. The first file contains all variables, whereas the second file contains only the *a* variable. Let us verify this results.

```
1  >> load('test1.mat')
2  >> whos
3    Name        Size           Bytes      Class
      Attributes
4
5    a           2x4              64        double
6    b           1x13             26        char
7    c           1x2              16        double
8  >> load('test2.mat')
9  >> whos
10   Name        Size           Bytes      Class
      Attributes
11
12   a           2x4              64        double
```

The default file type for saving workspace is .mat files; however, we can specify the file type by adding the required extension to the file name as follows:

```
1  >> xlswrite('test2.xls',a)
2  >> xlsread('test2.xls')
3
```

```
4  ans =
5
6      1      2      3      4
7     23     24     25     26
```

1.11 WORKSPACE USER INPUT

A useful function, *input*, waits for an input from the keyboard, ending with the ENTER key. You can try it out using the following example:

```
1  >> x=input('How  old  are  you? ')
2  How  old  are  you?  35
3
4  x =
5
6     35
```

By default, the input function takes numbers from the user; however, if we want to enter a string we can add 's' to the command:

```
1  >> x=input('What  is  your  name?  ','s')
2  What  is  your  name?  Adam
3  x =
4
5  Adam
```

Exercises:

- $3 < 2$

- $[1\ 2\ 5\ 3\ 0\ 3] \sim= 3$

- $0 == [0\ 0\ 0]$

CHAPTER 2

Photoplethysmogram Signals

This chapter gives an overview of photoplethysmogram (PPG) signals.

2.1 LEARNING OBJECTIVES

The learning objectives of this chapter are to:

- Learn about the mechanics of photoplethysmography and understand how it works

- Develop an understanding of the challenges around collecting the photoplethysmogram signal via pulse oximetry

- Gain insight into how photoplethysmogram signals collected via pulse oximetry can help in detecting abnormalities

2.2 BACKGROUND

Each heartbeat in the body is accompanied by variations in blood flow and volume, and these changes are measured by a medical device known as a *plethysmograph*. In terms of its etymology, the word *plethysmograph* is derived from two ancient Greek words; *plethysmos*, defined as "increase", and *graph* defined as "write". Various types of plethysmographs exist, each of which measures changes in blood volume with specific transducers that can differ from device to device. Photoplethysmogram (PPG) signals are optically obtained plethysmograms and represent the volumetric measurement of an organ. PPGs are often obtained using a pulse oximeter, which is a medical device that illuminates the skin with a light emitting diode (LED) and then measures changes in the skin's light absorption.[1,2]

With each cardiac cycle, the heart pumps blood to the periphery of the body. Even though this pressure pulse is somewhat damped by the time it reaches the skin, it is enough to distend the arteries and arterioles in the subcutaneous tissue, thereby delivering a signal that can be measured.[2] Changes in volume caused by the pressure pulse can be detected using a device known as a pulse oximeter. With this device, changes in blood flow are detected by illuminating the skin with the light from an LED diode and then measuring the amount of light received by a photodiode (PD). Each cardiac cycle appears as a peak, as seen in the Figure 2.1. Because

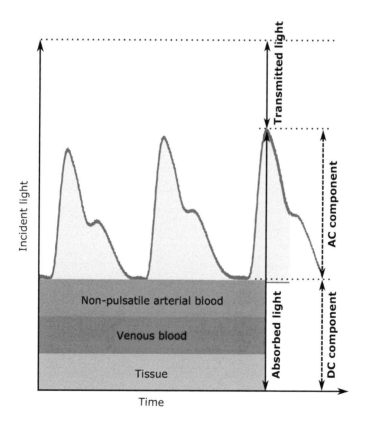

FIGURE 2.1 Light transmission and absorption within tissue.

blood flow to the skin can be modulated by multiple other physiological systems, the PPG can also be used to monitor breathing, hypovolemia, and other circulatory conditions.

2.3 OXYGEN TRANSPORT

Every cell in the human body depends on oxygen for survival. Blood in the body transports oxygen to these cells consistently. Within the blood, hemoglobin carries oxygen from the respiratory organ (i.e. lungs) to the rest of the body and tissues. When hemoglobin is saturated with oxygen molecules, it is referred to as oxyhemoglobin (red blood); desaturated hemoglobin molecules are referred to as deoxyhemoglobin (blue blood, clinically this is visible through the skin after periods of oxygen desaturation). Red blood absorbs more infrared light and allows more red light to pass than blue blood, which allows more infrared light to pass. This optical effect (i.e. light penetration) offers a straightforward and non-invasive way to measure the concentration of oxygen in the blood.

2.4 TERMINOLOGIES AND ACRONYMS

A photoplethysmogram is defined as a volumetric measurement of an organ and is a popular topic of study in engineering, science, medicine, and other related disciplines. Throughout the literature that discusses this topic, the term is often used interchangeably with other terminologies, such as "photoelectric plethysmography" and "photoplethysmography". Moreover, various acronyms are used to represent the term "photoplethysmogram signals" in the literature, such as PTG and DVP. Standard terminology and acronym usage is required to minimize the disconnect between researchers and disciplines, and to create knowledge cohesiveness. Moreover, it improves the searchability of all related publications in the field, facilitating an optimized and focused use of current data and information, therefore enriching the research process. The same is true for terms used to describe the second derivative of the photoplethysmogram. Evidence and examples are provided in Sections 2.4.1–2.4.7, which demonstrate inconsistencies in the current literature, and a recommendation is made for the scientific community.

2.4.1 DVP

The term "digital velocity pulse" (DVP) has been used by many researchers when referring to photoplethysmogram signals. Here are a few examples of publications using this terminology and the context in which the term was used.

- In 2002, Millasseau et al.[3] used DVP terminology when providing a simple, reproducible, non-invasive measure of large artery stiffness using the contour analysis of the DVP.

- Padilla et al.[4] used DVP terminology when suggesting a useful relation between the stiffness index $SI_{DVP} = h/\Delta T_{DVP}$ obtained in a simple form by means of DVP and the ankle-brachial pulse wave velocity (abPWV), as well as the values of sanguineous pressure, whereas abPWV = Lab/ΔT, and Lab measured the distance between the ankle-brachial location.

- In 2007, Alty et al.[5] used the DVP terminology when they developed a method with which to classify subjects accurately into high and low PWV (equivalent to high and low CVD risk) using features extracted from their DVP waveform.

2.4.2 PTG

The term "plethysmogram" (PTG) has been used by many researchers when referring to photoplethysmogram signals. Here are a few examples of publications using this terminology and the context in which the term was used.

- In 1999, Mishima et al.[6] used PTG terminology when they found that an increase in mean RR interval and a decrease in baseline deflection of the PTG strongly correlates with autogenic training.

- In 2005, Yashima et al.[7] used PTG terminology when they proposed a new stress evaluation technique using the photoplethysmogram (PTG) by applying Morlet wavelets.

- In 2007, Kageyama et al.[8] used PTG terminology when they performed wavelet analysis of fingertip photoplethysmogram (PTG) in order to quantify the stress stage.

- In 2008, Abe et al.[9] used PPG terminology when they proposed a method for evaluating effects of visually induced motion sickness.

- In 2008, Cox et al.[10] used PPG terminology when they proposed the use of photoplethysmogram (PPG) morphology as an indicator of hypovolemic states and their correlation with blood pressure.

- In 2009, Gil et al.[11] used PPG terminology when they analyzed heart rate variability (HRV) during decreases in the amplitude fluctuations of photopletysmography (PPG) events for obstructive sleep apnea syndrome screening.

2.4.3 SDPTG

The term "second derivative photoplethysmogram" (SDPTG) has been used by many researchers when referring to the second derivative of the photoplethysmogram. Here are a few examples of publications using this terminology and the context in which the term was used.

- In 1998, Takazawa et al.[12] used SDPTG terminology when they demonstrated that the b/a ratio reflects increased arterial stiffness, hence the b/a ratio increases with age. Moreover, they found that the $(b - c - d - e)/a$ ratio may be useful for evaluation of vascular aging and for screening of arteriosclerotic disease.

- In 1998, Imanaga et al.[13] used SDPTG terminology when they provided direct evidence that the magnitude of b/a of the APG is related to the distensibility of the peripheral artery, and suggest that the magnitude of b/a is a useful non-invasive index of atherosclerosis and altered arterial distensibility.

- In 2007, Baek et al.[14] used SDPTG terminology when they confirmed that the b/a ratio and the aging index (SDPTG-AI) increased with age, and the c/a, d/a, and e/a ratios decreased with age. The informal aging index $(b - e)/a$ has been suggested as a replacement for the known formula $(b - c - d - e)/a$.

- In 2007, Kimura et al.[15] used SDPTG terminology when they proposed the efficacy of the Kamishoyosan herb for patients with premenstrual syndrome was quantitatively ascertained using the second derivative of the photoplethysmogram signals.

2.4.4 APG

The term "acceleration photoplethysmogram" (APG) has been used by many researchers when referring to the second derivative of the photoplethysmogram. Here are a few examples of publications using this terminology and the context in which the term was used.

- In 1993, Katsuki et al.[16] used the APG terminology when they suggested that the present APG index is adequate about the individual reproducibility, and it has the possibility to apply as an index of arteriosclerosis.

- In 1995, Takada et al.[17] used the APG terminology when they concluded that simply categorized wave patterns of APG could be a useful non-invasive tool to evaluate aging in cardiovascular system.

- In 2000, Bortolotto et al.[18] used APG terminology when they introduced the APG Age Index as a useful measure for the evaluation of vascular aging in hypertensives.

- In 2005, Ushiroyama et al.[19] used APG terminology when they reported that patients with a sensation of coldness showed an improvement of the APG index $(b - c - d)/a$ following Sho treatment, which includes Kamishoyosan. Sho is one of the main concepts of Kampo medicine, and corresponds to the holistic tailored treatment suitable for the individual patient's symptoms.

- In 2006, Nousou et al.[20] used APG terminology when they developed a diagnosis assistance system with a light load to the testee. This system measures the acceleration plethysmogram and reproduces the diagnosis by using self-organizing maps.

- In 2007, Taniguchi et al.[21] used APG terminology when they proposed a method for using APG variability to evaluate a surgeon's stress when using a robotic surgical assistance system.

- In 2008, Fujimoto et al.[22] used APG terminology when they proposed the possibility of diagnosing stress through the analysis of an accelerated plethysmogram, which combines two evaluations based on chaos theory: the trajectory parallel measurement method, and the size of the neighboring space in the chaos attractor.

2.4.5 SDDVP

The term "second derivative digital velocity pulse" (SDDVP) has been used by many researchers when referring to the second derivative of the photoplethysmogram. Here are a few examples of publications using this terminology and the context in which the term was used.

- Millasseau et al.[23] used the term when they used the photoplethysmogram to examine the vascular impact of aging and vasoactive drugs.

- Rivas-Vilchis et al.[24] used the term when they used the photoplethysmogram to assess the vascular effects of PC6 (Neiguan) in healthy and hypertensive subjects.

2.4.6 Terminology Selection and Search Strategy

As mentioned, there are inconsistencies in the terminology and acronym usage for plethysmogram signals. For the purposes of this book, and as a suggested standard for any researchers in the area of photoplethysmogram signals, the terminologies *photoplethysmogram* (PPG) and the *second derivative of photoplethysmogram* (APG) signals will be used.

The following criteria were used during the selection process for terminology and acronyms:

1. Simplicity: The term should be as simple as possible.

2. Meaning: An acronym should meaningfully represent the nature of the term and the nature of its meaning; the acronym should feel intuitive (e.g. the first derivative of distance is velocity, not the FDD).

3. Repetition: The term is one most commonly used in the literature.

Criteria 1 and 2 can be subjective, whereas criterion 3 is quantitative and therefore empirical data is available from which to draw firm conclusions. Terminology and acronym data were collected from the PubMed database for English publications in the field of PPG study. Science Links Japan was also used during this process, as there is a large field of study dedicated to the photoplethysmogram in Japan. The PubMed database is comprised of more than 19 million citations for biomedical articles from MEDLINE and life science journals, going as far back as 1948, making it an ideal resource of the purpose of this literature review. Results of the literature review that show the number of uses for each term and acronym are as follows:

PubMed – *Terms*

- Photoplethysmography = 1,024
- Photoplethysmogram = 91

PubMed – *Acronyms*

- PPG = 232
- PTG = 13
- DVP = 10
- APG = 4
- SDPTG = 15

Science Links Japan – *Terms*

- Photoplethysmography = 23
- Photoplethysmogram = 24

Science Links Japan – *Acronyms*

- PPG = 2
- PTG = 4
- DVP = 8
- APG = 28
- SDPTG = 8

2.4.7 Standard Acronyms

Results show that *photoplethymogram* is the most commonly used term, and *PPG* is the acronym most commonly associated with photoplethysmogram) for all literature reviewed. For the second derivative, the most commonly used term is the *second derivative of photoplethysmogram* and the most commonly associated acronym is *SDPTG* in the English literature review. In Japanese literature, the most commonly used term for the second derivative is *acceleration photoplethysmogram* and the most commonly associated acronym is *APG*. Based on the outlined selection criteria, the term *photoplethymogram* and its associated acronym *PPG*, and the term *acceleration photoplethysmogram* and its associated acronym *APG* will be the terms and acronyms used as the standard. The scientific community is strongly advised to follow these recommendations to create consistency. The usage of different acronyms to describe the same signal can lead to a disconnect between different groups of engineers and scientists. Setting this standard will save researchers time and effort, and will overcome the generation of redundant results. It will also speed up scientific progress in this multidisciplinary field.

2.5 WHY PPG SIGNAL?

PPG signal data collection and analysis is easy to set up, and is convenient, simple, and economically efficient. Collection does not require direct contact with the skin surface, unlike other plethysmograph methods. Typically, a probe (recently, PPG has been measured via smartwatches and smartphones) is used that contains a light source and a detector to detect cardiovascular pulse waves that propagate through the body. For all these reasons, PPG is an attractive biosignal, especially for wearable applications.

The PPG signal reflects blood movement in a vessel: blood travels from the heart to the fingertips and toes through the blood vessels in a wave-like motion,[25] as shown in Figure 2.2 (PPG). An optical measurement technique is used, creating an invisible infrared light that is sent into the tissue. The amount of the backscattered light corresponds with the variation of the blood volume.[26] Hertzman was the first to find a relationship between the intensity of the backscattered light and blood volume in 1938.[27] The low-cost and simplistic manner of this optically based technology could offer significant benefits to health care (e.g. in primary care where non-invasive, accurate, and simple-to-use diagnostic techniques are desirable). Further, the development of PPG technologies and methodologies holds a great deal of promise in creating more accurate and effective tools for the management of not only vascular disease, but also other diseases.

As shown in Figure 2.2 (PPG), the wave contour of the PPG signal seems non-informative and simple in its appearance, and thus was not analyzed and investigated. Due to this simplicity, extracting information from the PPG signal is difficult, especially when trying to detect changes in the phase of the inflections. Ozawa[12] introduced the first and the

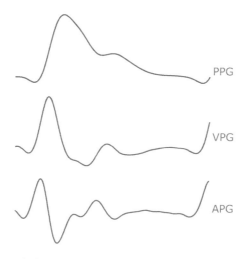

FIGURE 2.2 Photoplethysmogram signal and its derivatives. *Note*: PPG refers to the original fingertip photoplethysmogram; VPG refers to the first derivative of the photoplethysmogram; APG refers to the second derivative of the photoplethysmogram.

second derivatives of the PPG signal, as shown in Figure 2.2 (VPG and APG, respectively), to facilitate the interpretation of the original PPG waves in an attempt to extract information from the wave. The first and second derivatives of the PPG signal were developed as methods that more accurately recognized inflection points of the wave, thereby allowing for an easier interpretation of the original PPG wave. Recently, analysis of the PPG waveform has attracted increasing interest, especially in circulatory[28] and respiratory[29] monitoring.

2.6 PLETHYSMOGRAPHY TYPES

The general plethysmography types are: water, air, strain gauge, impedance, and photoelectric. The photoelectric plethysmogram—as we call it, the photoplethysmogram—is the most convenient and efficient type compared to others. Here is more information about the different types of plethysmography:

- Water
  - Water-filled cuff
  - Water-filled body
  - Water-filled chamber

- – Measuring penile blood flow
- – Measuring Pulmonary Capillary Blood Flow
- – Measuring maximal blood flow
- Air
  - Air-filled cuff
    - – Evaluation of venous hemodynamics
    - – Measures parameters of global venous function such as calf venous volume, venous filling index, ejection fraction, and residual volume fraction
- Strain gauge
  - Fine rubber tube (filled with mercury)
    - – Assessment of capillary filtration
    - – Assessment of volume changes in venous diseases
    - – Identifying limbs with suspected venous incompetence
    - – Evaluation of peripheral circulation in spinal cord injury cases
    - – Evaluation of acute and chronic venous insufficiency
    - – Evaluation of peripheral vascular disease
    - – Measurement of deep venous thromboses
- Impedance
  - Electrodes
    - – Detection of blood flow disorders
    - – Assessment of fat-free mass of the human body
- Photoelectric
  - Photodetectors
    - – Monitoring of heart and respiratory rates
    - – Monitoring of oxygen saturation

- Assessment of blood vessel viscosity

- Assessment of venous function

- Measuring the ankle pressure

- Measuring genital responses

- Assessment of venous reflux

- Measuring cold sensitivity

- Measuring blood pressure

- Assessment of cardiac output

2.7 MEASURING SITES

The signal collected from the fingertip is affected by the heartbeat, the hemodynamics and the physiological condition caused by the change in the properties of an arteriole, as seen in Section 5.3. These effects can be observed as distortions in the wave profiles. The location of the LED and PD is an important design element that affects the signal quality and robustness from motion artifacts. Therefore, suitable measurement sites must be located to optimize sensor performance. PPG sensors are commonly worn on the fingers due to the high signal amplitude that can be achieved in comparison to other sites.[27] However, this configuration is not well-suited to pervasive sensing, as most daily activities involve the use of the fingers. In recent years, different measurement sites for PPG sensors have been explored extensively, including the ring finger,[30] wrist,[31] brachia,[32] earlobe,[33] external ear cartilage,[34] and the superior auricular region.[35] In addition, the esophageal region has been used in clinical practice.[36] Commercial clinical PPG sensors commonly use the finger, earlobe, and forehead.[37] Use of a glass-type wireless PPG has also been examined.[38]

Perfusion values (values that report on pulse strength) of 52 anatomical sites in healthy subjects showed that the fingers, palm, face, and ears offer much higher perfusion values when compared to other measurement sites;[39] transmitted PPG signal amplitude from the earlobe provides the largest perfusion value. In addition, earlobe sensors are easy to fabricate and have become popular as pulse rate monitors. However, a spring-loaded ear-clip, although effective, can become painful over extended monitoring periods. There was little improvement in the wearable earlobe PPG sensor design until the development of micro-electromechanical system (MEMS) technology. MEMS facilitated the fabrication of a lightweight, comfortable, fully integrated, self-contained sensor earpiece. For example, an earring PPG sensor with a magnetic attachment to the earlobe was developed that allowed good contact for monitoring during physical activity.

2.8 MODES OF PPG MEASUREMENT

Pulse oximetry is based on the principle of red light and infrared light absorption characteristics of oxygenated and deoxygenated hemoglobin. Oxygenated hemoglobin absorbs more infrared light and allows more red light to pass through. Deoxygenated (or reduced) hemoglobin absorbs more red light and allows more infrared light to pass through. Red light is in the 600–750 nm wavelength light band, whereas infrared light is in the 850–950 nm wavelength light band, as shown in Figure 2.3. Pulse oximetry uses a light emitter comprising red and infrared LEDs that shines through a reasonably translucent site with good blood flow. Opposite to

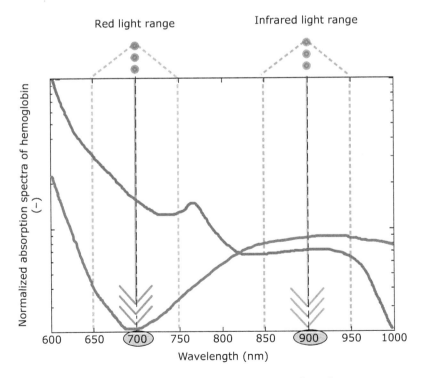

FIGURE 2.3 Absorption spectra of hemoglobin. *Note*: The red curve represents oxygenated hemoglobin, while the purple curve represents the deoxygenated hemoglobin.

the emitter is a photodetector that receives the light that passes through the measuring site. Typical adult sites are the finger, toe, pinna (top), or ear lobe. Typical infant sites include the foot or palm of the hand, and the big toe or thumb. Light is sent through the measuring site by two methods: *transmissive* and *reflective.*

Light from the red and infrared LEDs is absorbed by several parts of the body at the measurement site, which will be referred to as *light absorbers.* These components are the skin, tissue, venous blood, and the arterial blood. However, with each heartbeat the heart contracts, resulting in a momentary surge of arterial blood flow, which consequently increases arterial blood volume across the measuring site. More light is absorbed during the surge. When viewing the light signals received at the photodetector in the shape of a waveform, there are apparent peaks with each heartbeat and apparent troughs between heartbeats. If the light absorption at the trough (which includes all the light absorbers) is subtracted from the light absorption at the peak then, in theory, the resulting value reflects the absorption characteristic of the added blood, and is considered the arterial blood flow. Many problems inherent to oximetry measurement in the past were overcome with this new method, which is the conventional method used today. Interestingly, the term "pulse oximetry" was coined due to the measured peaks occurring with each heartbeat or pulse.

Despite these developments in measurement, conventional methods still struggle with accuracy, especially during motion and low perfusion. Arterial blood gas tests have been, and continue to be commonly used to supplement or validate pulse oximeter readings. The advent of "Next Generation" pulse oximetry technology has demonstrated significant improvement in the ability to read through motion and low perfusion; thus making pulse oximetry a more dependable method on which to base medical decisions.

2.8.1 Transmissive Mode

In the transmissive method, as shown in the Figure 2.4(a), the emitter and photodetector are opposite each other, the measurement site being in between. The light can then pass through the site. In the reflectance method, the emitter and photodetector are next to each other on top of the measuring site. The light bounces from the emitter to the detector across the site. Transmission is the method most commonly used, and all examples and discussions related to pulse oximetry throughout this text have used this preferred method as the underlying approach.

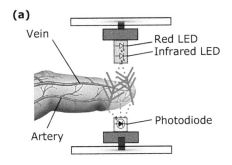

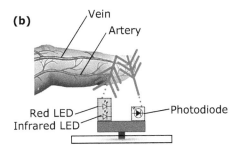

FIGURE 2.4 Modes of PPG measurement: (a) transmissive mode, (b) reflective mode.

2.8.2 Reflective Mode

Wearable PPG devices have two modes during signal collection, as shown in Figure 2.4(b). In transmission mode, the light transmitted through the medium is detected by a PD opposite the LED source. In reflectance mode, the PD detects light that is back-scattered or reflected from tissue, bone, and/or blood vessels. The transmission mode is capable of obtaining a relatively good signal, but the measurement site may be limited. To be effective, the sensor must be located on the body at a site where the transmitted light can be readily detected, such as the fingertip, nasal septum, cheek, tongue, or earlobe. Sensor placement on the nasal septum, cheek, or tongue is only effective under anesthesia. The fingertip and earlobe are the preferred monitoring sites; however, these sites have limited blood perfusion. In addition, the fingertip and earlobe are more susceptible to environmental extremes, such as low ambient temperatures (e.g. for military personnel or athletes in training). The greatest disadvantage is that the fingertip sensor interferes with daily activities.

The reflectance mode eliminates the problems associated with sensor placement during transmission mode, and a variety of measurement sites can be used. However, the reflection mode is affected by motion artifacts and pressure disturbances. Any movement, such as physical activity, may lead to motion artifacts that corrupt the PPG signal and that limit the measurement accuracy of physiological parameters. Pressure disturbances acting on the probe, such as the contact force between the PPG sensor and the measurement site, can deform the arterial geometry by compression. Thus, in the reflected PPG signal, the AC amplitude may be influenced by the pressure exerted on the skin. Reflectance PPG sensors such as the MaxFast (Nellcor™, Mansfield, MA, USA) have been used clinically for the non-invasive continuous measurement of oxygen saturation. Anecdotally, it has been reported that this measurement type can occasionally yield false positive readings; however, further research is needed in this area.

2.9 CALCULATION OF OXYGEN SATURATION

Regardless of the mode of measurement, whether transmission or reflective (as discussed in Section 2.8), once the red (R) and infrared (IR) signals are received at the photodetector, the R/IR ratio is calculated. The R/IR is compared to a "look-up" table (comprising empirical formulas) that convert the ratio to an SpO2 value. Most manufacturers have their own look-up tables based on calibration curves derived from healthy subjects at various SpO2 levels. Typically, an R/IR ratio of 0.5 equates to approximately 100% SpO2, a ratio of 1.0 to (approximately) 82% SpO2, and a ratio of 2.0 equates to 0% SpO2. The major change since the 8-wavelength Hewlett Packard oximeters of the 1970s is that the oximeters of today include arterial pulsation to differentiate the light absorption in the measurement site due to skin, tissue, and venous blood from the arterial blood. Note that only one LED is required to generate a PPG signal.

2.10 SIMULATION OF PPG SIGNAL USING SINUSOIDS

We can simply simulate PPG signals using sinusoids. First, let us see whether we should use a sine wave or cosine wave, as follows:

```
1  >> x=0:pi/100:2*pi;
2
3  >> wave_1=cos(x*2);
4  >> figure , plot(wave_1)
5  >> xlabel("Angle")
```

```
6 >> ylabel("Amplitude")
7 >> title ("An example of generating a waveform using
     a sinusoid");
```

We generated a vector of x-axis data using the variable x from 0 to 2π in steps of $\pi/100$. The first waveform was generated using one sinusoid (or cosinusoid). Note that the *cos* function operates element-wise on arrays. As we can see in Figure 2.5, one sinusoid is insufficient to simulate the PPG waveform. Note the two main waves in the PPG waveform generated by systole and diastole.

Therefore, it is expected that we at least need two sinusoids to simulate the two phases of the cardiac cycle. The following code shows how more than one sinusoid can formulate a waveform similar to that of a PPG:

```
1 >> x=0:pi/100:2*pi;
2 >> wave_2=cos(x*3)+cos(x*7-2);
3 >> figure , plot(wave_2)
4 >> xlabel("Angle")
5 >> ylabel("Amplitude")
6 >> title ("How to simulate PPG waveforms using
     sinusoids in Matlab.");
```

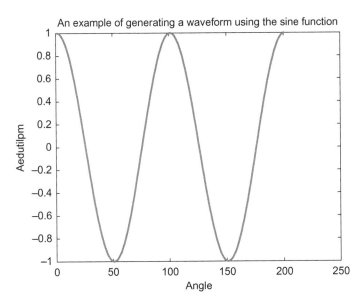

FIGURE 2.5 Simple example of how to plot a sinusoid signal in MATLAB.

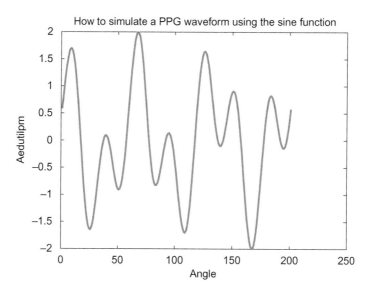

FIGURE 2.6 Simple example of how to simulate a PPG waveform using sinusoids.

In Figure 2.6, we can see the simulated systolic and diastolic waves in a PPG signal.

Note that the first sinusoid represents *systole*, which occurs when the heart contracts to pump blood out, and the sinusoid representing *diastole* occurs when the heart relaxes after contraction. This simulation is not completely accurate; therefore, Section 2.11 will discuss how to improve PPG simulation using Gaussian functions.

2.11 SIMULATION OF PPG SIGNAL USING TWO GAUSSIAN FUNCTIONS

We can simulate one PPG pulse waveform using Gaussian functions. First, let us see the waveform of the Gaussian function.

```
1 >> a = 1;
2 >> mu = 50;
3 >> sigma = 10;
4 >> x = 1:1:100;
5 >> y = a * exp(-(((x-mu)/sigma).^2)/2) ;
6 >> figure; plot(x ,y);
7 >> xlabel("Sampling points");
8 >> ylabel("Amplitude");
9 >> title("An example of generating a waveform using
     one Gaussian function");
```

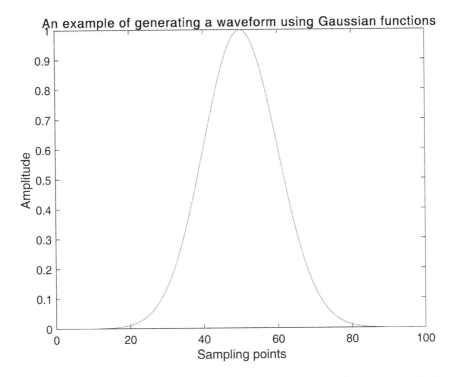

An example of generating a waveform using Gaussian functions

FIGURE 2.7 Simple example of how to plot a Gaussian function signal in MATLAB.

We generated a vector of x-axis data using the variable x from 1 to 100 in steps of 1, which comprise the sampling points of one pulse. The first waveform was generated using one Gaussian function. The parameters a, mu, and sigma are, respectively, the height of the peak, the center of the peak and the standard deviation of the Gaussian function. Note that the Gaussian function operates element-wise on arrays. As we can see in Figure 2.7, one Gaussian function is insufficient to simulate the PPG waveform. Note the two main waves in the PPG waveform generated by systole and diastole. Therefore, it is expected that we need at least two Gaussian functions to simulate the two phases of the cardiac cycle. The following code shows how more than one Gaussian function can formulate a waveform similar to that of a PPG:

```
1 >> a = [0.8,0.4];
2 >> mu = [25,50];
3 >> >> sigma = [10,20];
4
5 >> x = 1:1:100;
6 >> y1 = a(1) * exp(-(((x-mu(1))/sigma(1)).^2)/2);
```

```
 7 >> y2 = a(2) * exp(-(((x-mu(2))/sigma(2)).^2)/2);
 8 >> y = y1 + y2;
 9
10 >> figure; plot(x,y,'b');
11 >> hold on; plot(x,y1,'k—');plot(x,y2,'r—');
12 >> xlabel("Sampling points");
13 >> ylabel("Amplitude");
14 >> legend("Synthetic PPG","1^{st} Gaussian",
      "2^{nd} Gaussian");
15 >> title("An example of generating a waveform using
      two Gaussian functions");
```

In Figure 2.8, we can see the simulated phases of one cardiac activity in PPG waveforms. Note that the different values of parameters will obtain different waveforms.

Because the length and waveform of pulses vary, the parameters vary from pulse to pulse. If we ignore the inter-beat variation of the waveform,

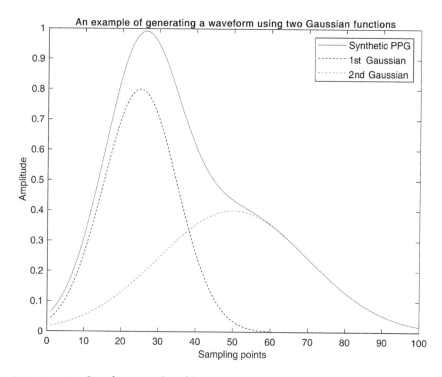

FIGURE 2.8 Simple example of how to simulate PPG waveform using two Gaussian functions.

we can simply use the same parameters in different pulses, synthesize the pulse to a fixed length, then use upsampling or downsampling to the required length. Another method is to use a circular motion to represent the periodicity of PPG. One cycle of movement on the circle corresponds to a PPG pulse. The independent variable of Gaussian functions is replaced by an angle in the range $[-\pi, \pi]$. The number of sampling points in one beat is related to the angle velocity.

The following code shows the generation of one PPG waveform after adding a circular motion (i.e. introducing angle range for periodicity) to two Gaussian functions model:

```
1  >> Duration = 1;
2  >> Fs = 125; %Sampling Frequency
3
4  >> a = [0.82,0.4];
5  >> mu = [-pi/2,0];
6  >> sigma = [0.6,1.2];
7
8  >> Samples = Fs/Duration;
9  >> V_angle = 2*pi/Samples;
10 >> angle = -pi+V_angle :V_angle:pi;
11 >> y1 = a(1) * exp(-(((angle-mu(1))/sigma(1)).^2)/2);
12 >> y2 = a(2) * exp(-(((angle-mu(2))/sigma(2)).^2)/2);
13 >> y = y1 + y2;
14
15 >> figure; plot(angle,y,'b');
16 >> hold on; plot(angle,y1,'k-');plot(angle,y2,'r-');
17 >> xlabel("Angle");
18 >> ylabel("Amplitude");
19 >> xlim([-pi,pi]);
20 >> set(gca,'xtick',[-pi,0,pi],'xticklabel',{'\pi',
      '0','\pi'});
21 >> legend("Synthetic PPG","1^{st} Gaussian", "2^{nd}
      Gaussian");
22 >> title("An example of generating a waveform using
      angle model");
```

Figure 2.9 shows the synthesized PPG waveform based on periodicity. Connecting individual PPG, the pulses generated produce a long PPG recording.[40,41]

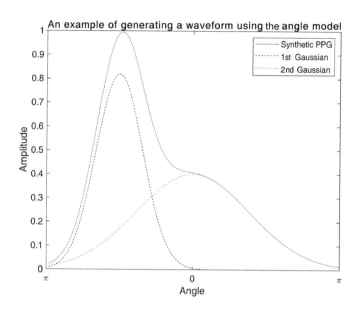

FIGURE 2.9   An example of a synthesized PPG waveform using two Gaussian functions after adding periodicity to the model.

2.12 PPG SENSORS

2.12.1 Probe-Based PPG Signals

Typically, the probe-based PPG sensor consists of two parts: light emitting diodes (LEDs), and a light detector (photodetector). Beams of light are shone through the tissues from one side of the probe to the other. The blood and tissues absorb some of the light emitted by the probe. The light absorbed by the blood varies with the oxygen saturation of hemoglobin. The photodetector detects the light transmitted as the blood pulses through the tissues and the signal generated is called photoplethysmogram (PPG).

In order for the PPG sensor to function properly, the probe must be placed where a pulse can be detected. The LEDs must face the light detector in order to detect the light as it passes through the tissues. The probe emits a red light when the machine is switched on. Probes are designed for use on the finger, toe, or ear lobe. They are four main PPG probes, as shown in Figure 2.10. Hinged probes are the most popular, but are easily damaged; rubber probes are the most robust. The wrap around design of some probes may constrict the blood flow through the finger if put on too tightly. Ear probes are lightweight and are useful for use with children, or when a patient is very vasoconstricted. Small probes have been designed for use with children, but an adult hinged probe may be used on the thumb

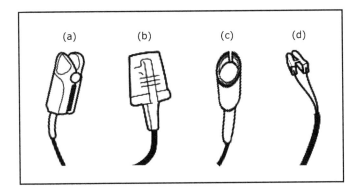

FIGURE 2.10 Types of probe-based PPG sensors: (a) hinged finger probe, (b) rubber finger probe, (c) wrap finger probe, (d) clip ear probe.

or big toe of a child. For finger or toe probes, the manufacturer marks the correct orientation of the nail bed on the probe.

The probe is highly delicate and can easily be damaged; it must therefore be handled with care. Any movement to the probe or its cable will impact the quality of the recorded PPG signal.

2.12.2 Video-Based PPG Signals

Photodiodes can be used in the form of a video camera, which is termed *video plethysmography*. Two types of video plethysmography are used: non-contact (also referred to as "Remote PPG") and contact (also referred to as "Imaging PPG" or "iPPG"). Non-contact is used for remote measurements, and uses a charge coupled device (CCD) camera, such as a mobile phone camera or web camera, and examines the light reflected from skin. Plethysmographic signals can be measured remotely (>1 m) using ambient light and a PD with either a simple consumer-level digital camera in movie mode or a cellular phone.[42–43] Although the green channel features the strongest plethysmographic signal, corresponding to an absorption peak by (oxy-)hemoglobin, the red and blue channels also contain plethysmographic information. These results suggest that ambient light PPG may be useful for medical purposes, such as the characterization of vascular skin lesions, or the remote sensing of vital signs for triage or sports purposes.

The contact system is used for close measurements, such as placing the finger on the mobile phone camera to cover the entire camera view and using the phone camera flash LED light (white light) to illuminate the finger.[45] This type of PPG measurement method has been commercialized by Azumio (http://www.azumio.com/apps/heart-rate/).

2.13 CURRENT CHALLENGES

There are many challenges to the generation of a reliable PPG application for assessing a specific abnormality. These challenges, such as several types of additive artifacts, affect not only the signal processing steps (e.g. extraction of features), but also the overall diagnosis. The main PPG challenges are powerline interference, a sudden change in amplitude, motion artifacts, multi-parameter systems, and research design.

2.13.1 Powerline Interference

This type of artifact is associated with instrumentation amplifiers, where the recording system picks up ambient electromagnetic signals and other artifacts. High frequency artifacts caused by mains power source interference is induced onto the PPG recording probe or cable. This artifact introduces a sinusoidal component into the recording. In Australia, this component is at a frequency of 50 Hz. The periodic interference is clearly displayed as a spike in Figure 2.11(a) not only at its fundamental frequency of 50 Hz, but also as spikes at 100 Hz and higher harmonics.

2.13.2 Sudden Amplitude Change

This type of artifact occurs when the automatic gain controller adjusts the gain of the amplifier automatically based on the amplitude of the input signal. This may cause amplitude saturation in the amplitude of the PPG waveform at a maximum or minimum value, or may cause the signal to rest at some random fixed value. A low amplitude PPG signal caused by the automatic gain controller is shown in Figure 2.11(b). However, the reduction of PPG amplitude can be directly attributable either to a loss of central blood pressure, or to constriction of the arterioles perfusing the skin.

2.13.3 Motion Artifact

This type of artifact is likely caused by poor fingertip contact with the photo sensor. Variations in temperature and bias in the instrumentation amplifiers can sometimes cause baseline drift, too. It is difficult to set up measurement procedures that yield PPG signals without low frequency artifacts. Figure 2.11(c) shows a noisy PPG signal with powerline and motion artifacts. The low frequency artifact can be removed using a high pass filter, or vice versa. Usually, the cause of motion artifacts is assumed to be due to vibrations or movement of the subject. The shape of the baseline disturbance caused by motion artifacts can be assumed to be a biphasic signal resembling one cycle of a sine wave, as shown in Figure 2.11(d).

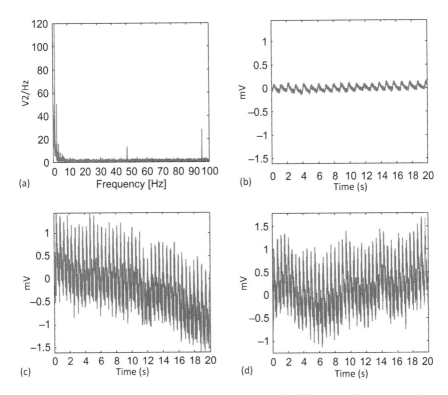

FIGURE 2.11   Challenges in analyzing PPG signals: (a) mains electricity noise, (b) low amplitude PPG signals, (c) powerline and motion artifacts in PPG signal, (d) baseline wandering in PPG signal.

2.13.4 Multi-Parameter Systems

It goes without saying that the greater the number of biosignals collected simultaneously, the more accurate the diagnosis will be. For example, researchers are combining PPG with one or more additional cardiac signals in order to generate a more accurate detection, estimation, or prediction of abnormality. For instance, many researchers have adopted several different physiological signals in the estimation of blood pressure, such as those from electrocardiography (ECG), ballistocardiography (BCG), and phonocardiography (PCG), in addition to PPG signals.[46]

One important issue with regard to collecting multiple biosignals simultaneously is synchronization. For example, it was assumed that biosignals in the MIMIC database were collected simultaneously and that the signals were also synchronized. Many papers published results under this assumption. There is a need for publicly available physiological databases that contain time-synchronized physiological signals so that researchers can examine time-based features adequately and find the most consistent and reliable feature that correlates with the abnormality being investigated.

2.13.5 Research Design

The research study design plays a major role in the findings, and there are many challenging steps that need more attention.

1. *Sample size*: There is a lack of power calculation step before running the study. Sample sizes need to comprise at least more than 100 subjects, with a mixture of both normal and abnormal subjects.

2. *Socioeconomic diversity*: There is a great need for diversity in age, race, gender, and so on.

3. *Testing condition*: The robustness of the developed algorithms needs to be tested under a variety of movement conditions, very cold and hot extremities, rather than simply taking sedentary measurements.

4. *Location of PPG measurement*: There is a need to optimize the locations for PPG measurement; for example, the forehead versus the toe (note that variability between different anatomical sites was reported by Allen and Murray)[47]).

5. *Inclusion of co-morbidities*: Most of the studies do not discuss co-morbidities, which could change the impact of a study's findings.

2.14 SUMMARY

In summary, the challenges discussed can exist simultaneously within a PPG signal, as shown in Figure 2.11. A PPG signal is complex and sensitive to artifacts; it requires patient and accurate analysis in order to retrieve the detailed information contained within the signal. It may be for these reasons that the PPG signal has not been widely investigated beyond its use in oximetry.[48] The challenges discussed are important to research groups around the globe, as resolving these issues will enable such groups to intensify their efforts toward creating a more accurate and robust PPG-based technology that is non-invasive, cuff-less, continuous, and calibration-free, and that can be easily integrated into wearable devices.

CHAPTER 3

Visualization of PPG Signals

This chapter discusses PPG signals' graphic representation to produce images that capture meaningful information (or features) from the raw PPG signals. Mainly, it is the process of mapping the raw PPG signal into a different domain (time, frequency, or time-frequency) to identify patterns associated with a specific abnormality.

3.1 LEARNING OBJECTIVES

The learning objectives of this chapter are to enable the:

- Visualization of raw biosignals into meaningful graphical representations

- Visualization and analysis of one-dimensional PPG signals using the standard periodogram, spectrogram, and/or wavelet methods

- Extraction of temporal-spatial information in order to gain scientific insight into the physical measurement of the signal

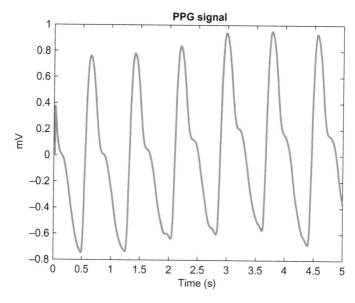

FIGURE 3.1 Data visualization using the *plot* function.

3.2 PLOT

A plot is a data visual technique used to see the PPG signal, usually as a graph showing the relationship between two or more variables. In our case, the plot data visual shows the relationship beween time and amplitude, as shown in Figure 3.1. The general MATLAB function for visualizing a PPG signal using plotting is:

```
1 >> plot(Y); %plots the columns of Y versus their index.
```

Here is a specific example for a PPG signal:

```
1 >> plot(PPG);
```

3.3 BAR

By definition, the bar graph is a graph that presents categorical data with rectangular bars with heights or lengths proportional to the values that they represent. However, as shown in Figure 3.2, when applied to a PPG signal (or any biosignal) the result is similar to the area graph. The general MATLAB function for visualizing PPG signals via a bar graph is:

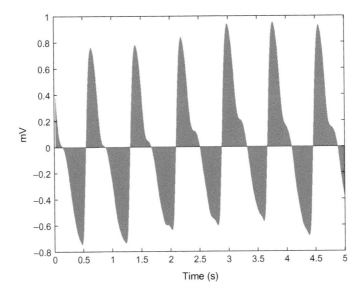

FIGURE 3.2   Data visualization using the *bar* function for a PPG signal as shown in Figure 3.1.

```
1  >> area(Y); %plots the vector Y or plots each column
      in matrix Y as a
2              %separate curve and stacks the curves.
```

Here is a specific example for a PPG signal:

```
1  >> area(PPG);
```

3.4 AREA

An area graph displays a PPG signal graphically and quantitatively by filling the area under the waveforms, as shown in Figure 3.3. The area between the axis and the line are commonly emphasized with colors, textures, and hatchings. It is almost identical to a bar graph. The general MATLAB function for visualizing PPG signals via an area graph is:

```
1  >> area(Y); %plots the vector Y or plots each column
      in matrix Y as a
2              %separate curve and stacks the curves.
```

Here is a specific example for a PPG signal:

```
1  >> area(PPG);
```

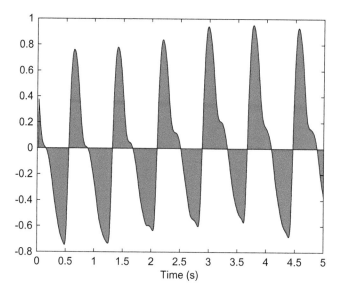

FIGURE 3.3 Data visualization using the *area* function for a PPG signal as shown in Figure 3.1.

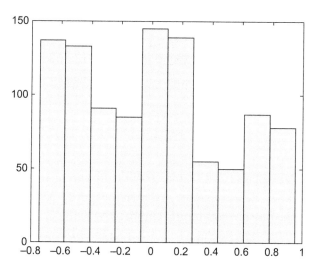

FIGURE 3.4 Data visualization using the *hist* function for the PPG signal shown in Figure 3.1.

3.4.1 Histogram

A histogram is a data visual that shows the distribution of the numerical data contained in a PPG signal, as shown in Figure 3.4. It is an estimate of the probability distribution of a continuous variable (quantitative variable) and was first introduced by Karl Pearson. The general MATLAB function for a histogram is:

```
1  N = hist(Y)  % bins the elements of Y into 10 equally
                   spaced containers
2         %and returns the number of elements in each
                   container. If Y is a
3         %matrix, hist works down the columns.
```

Here is a specific example for a PPG signal:

```
1  >> hist(PPG)
```

3.5 PERIODOGRAM

Periodograms, spectrograms, wavelets, and eventograms are used to evaluate, assess performance, and assess applicability. A periodogram views the processed biomedical signal as a sum of cosine waves with varying amplitudes and frequencies. A periodogram is obtained via the Fast Fourier Transform (FFT) and, typically, the frequency is displayed as the x axis, and the magnitude of the spectrum is displayed as the y axis. It is designed to identify important frequencies (periods) in biomedical signals; a periodogram graph displays the dominant frequency behavior within the signal. The periodogram, $R(f)$, is calculated as follows:

$$R(f) = \frac{1}{N}\left|\sum_{n=0}^{N-1} x_n e^{-i2\pi nf}\right|^2, \tag{3.1}$$

where x is the biomedical signal with length N, n is the data sample, and f is the frequency.

The basic MATLAB function for a spectrogram is:

```
1  Pxx = periodogram(X);  % returns the PSD estimate,
       Pxx, of a signal, X.
```

3.6 SPECTROGRAM

The methods for a spectrogram and a periodogram differ in that a spectrogram considers time localization (i.e. time–frequency visualization), whereas a periodogram only uses the frequency domain. A spectrogram is three-dimensional and includes the time, frequency, and magnitude of the spectrum; it comprises multiple short periods of spectrum combined together. The spectrogram is typically displayed with time as the x axis and frequency as the y axis. Colors are also used to show the magnitude of the spectrum. A spectrogram is typically obtained via the Short-Time Fourier Transform (STFT), which includes a windowing function to extract the local time signal before conducting the STFT. A spectrogram is designed to identify the dominant frequencies with respect to time in the processed biomedical signal. A spectrogram, $S(t,f)$, is calculated as follows:

$$S(t,f) = \left| \int_{-\infty}^{\infty} x(\tau)W(\tau-t)e^{jf\tau}d\tau \right|^2 ,$$

(3.2)

where x is the biomedical signal, $W(\tau)$ is the window function, and f is the frequency. The time index, t, is normally considered to be "slow" time and is not usually expressed in such high resolution as time τ. The basic MATLAB function for a spectrogram is:

```
1  P = pspectrum(X); %returns the power spectrum of X.
```

3.6.1 Wavelets

The wavelet is comparable to the spectrogram; however, the wavelet transform offers a multiresolution whereas the spectrogram only offers a fixed resolution. Wavelets are closely related to filter banks. The wavelet transform (WT) of the biomedical signal $x(t)$ is an integral transform defined by:

$$D(a,b) = \int_{-\infty}^{\infty} x(t)\psi_{a,b}^*(t)\ dt,$$

(3.3)

where $D(a, b)$ is known as the wavelet detail coefficient at the scale and location indices (a, b) of the biomedical signal $x(t)$, and $\psi*(t)$ denotes the complex conjugate of the wavelet function $\psi(t)$. The transform yields a *time-scale* representation similar to the *time–frequency* representation of the STFT, the WT uses a set of analyzing functions that allows a variable time and frequency resolution for different frequency bands. The set of analyzing functions—the wavelet family $\psi_{a,b}(t)$—is deduced from a *mother wavelet* $\psi(t)$ by:

$$\psi_{a,b}(t) = \frac{1}{\sqrt{2}}\psi\left(\frac{t-b}{a}\right),$$

(3.4)

where *a* and *b* are the *dilation* (scale) and *translation* parameters, respectively. The scale parameter a of the WT is comparable to the frequency parameter of the STFT. The mother wavelet is a short oscillation with a zero mean. The orthonormal dyadic discrete wavelets are associated with scaling functions and their dilation equations. The scaling function is associated with the smoothing of the signal and has the same form as the wavelet, given by $\phi_{a,b}(t) = 2^{-a/2}\phi(2^{-a}t - b)$. However, the convolution of the biomedical signal with the scaling function produces approximation coefficients as follows:

$$A(a,b) = \int_{-\infty}^{\infty} x(t)\phi_{a,b}(t)\ dt, \tag{3.5}$$

where the biomedical signal $x(t)$ can then be represented by a combined series expansion, using both the approximation coefficients A and the detail coefficients D. The basic MATLAB function for continuous wavelet is:

```
1 WT = cwt(X) %returns the continuous wavelet transform
  (cwt) of X.
```

3.7 EVENTOGRAM

Two moving averages are the basis of the novel visual=representation algorithm, termed the *eventogram*. The use of two moving averages has been previously tested on different biomedical signals for detecting *a* waves in APG (acceleration of PPG) signals,[49–51] for detecting *c*, *d*, and *e* waves in APG,[52] for detecting systolic waves in PPG signals,[53] for detecting QRS complexes in ECG signals,[54,55] for detecting T waves in ECG signals,[56] and for detecting heart sounds.[57] The approach using two moving averages involves generating the blocks of interest, which are then used to produce the *eventogram*. Additionally, power calculation, the rejection of noisy events, and image display are added as additional steps to produce the *eventogram*; the flowchart is shown Figure 3.5.

Generating blocks of interest: Blocks of interest are generated using two event-related moving averages that demarcate the systolic and heartbeat areas. In this procedure, the first moving average (MA$_1$) is used to emphasize the first event and is given by:

$$\text{MA}_1[n] = \frac{1}{W_1}(y[n-(W_1-1)/2]+...+ y[n]+... + y[n+(W_1-1)/2]), \tag{3.6}$$

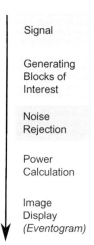

Signal

Generating
Blocks of
Interest

Noise
Rejection

Power
Calculation

Image
Display
(Eventogram)

FIGURE 3.5 Flowchart for generating an *eventogram*. This figure is adopted from.[58] *Notes: Eventogram* generation consists of four stages: feature extraction (generating potential blocks using two moving averages), the rejection of noisy events, power calculation, and image display.

where W_1 represents the window size of the duration of the first event. The resulting value is rounded to the nearest odd integer. The implementation in MATLAB can be as follows:

```
1  MA1 = smooth(PPG, W1);
```

The second moving average (MA_2) is used to emphasize the second event area to be used as a threshold for the first moving average and is given by:

$$MA_2[n] = \frac{1}{W_2}(y[n-(W_2-1)/2]+\ldots+y[n]+\ldots$$
$$+y[n+(W_2-1)/2]), \qquad (3.7)$$

where W_2 represents the window size of the duration of approximately one beat. Its value is rounded to the nearest odd integer. The implementation in MATLAB can be as follows:

```
1  MA2 = smooth(PPG, W2);
```

Here, blocks of interest are generated by comparing the MA_1 signal with MA_2. If any sample in $MA_1 \leq MA_2$, it will be replaced by zero. Samples that achieved this rule $MA_1 \leq MA_2$, will be replaced by zeros. Otherwise, it will be replaced by a value of one. The generated time series of a binary string is called "blocks of interest". However, some blocks of interest will contain noise segments that occur multiple times. The data within a series of ones are referred to as a B.

After obtaining the blocks of interests, block rejection of those that contain noise, calculating the power of each block, and then visualizing the power of the blocks in terms of W_1 and W_2, takes place as follows:

Noise rejection: Rejection is based on the average value of the standard deviations of all blocks. The standard deviation of each block (m) is calculated as follows:

$$\sigma_m = \sqrt{\frac{1}{L}\sum_{n=1}^{L}\left(B_m[n]-\bar{B}_m\right)^2},\qquad(3.8)$$

where L is the length of the data within each data block B.

After calculating the standard deviations of all blocks, a threshold (THR) is applied to reject blocks that contain noise as follows:

$$\text{THR} = \frac{1}{M}\sum_{m=1}^{M}\sigma_m,\qquad(3.9)$$

where M is the number of data blocks and THR is the average value of the standard deviations of all blocks. Usually, the noisy segments have a small standard deviation compared to the segments that contain informative events (i.e. wave, spike, or peak). Thus, the block that satisfies $\sigma > \text{THR}$ is accepted; otherwise, it is rejected.

Power calculation: The power of each block, calculated at a specific window length W_1 and a specific window length W_2, is used to represent a value in the *eventogram*, as follows:

$$P\left(W_1,W_2\right)=\sum_{m=1}^{M}B_m^2,\qquad(3.10)$$

where B is the data segment within a specific block m. This step generates a power matrix: $P_{u\times v}$, where u is the different values of (W_1) and v is the number of values of (W_2).

Image display: The P matrix (obtained in the previous step) is shown as an image with a range of colors. Each element of P specifies the color for one pixel of the image representing the intensity of the power over the data block generated using W_1 and W_2. Here, the *jet* range is used from blue to red, and passes through the colors cyan, yellow, and orange. The blue color refers to the lowest value of power and the red color refers to the highest value of power. The resulting image is a u-by-v grid of pixels, where u is the number of columns W_1 and v is the number of rows W_2 in P. The visual representation of this step is called an *eventogram*—the image display step produces the *eventogram*. The main goal of the *eventogram* is to identify the dominant time–domain events and their associated signal morphologies within the processed biomedical signal using W_1 and W_2. The pixel associated with the highest value of power is considered to contain the main event. The *eventogram* is a time–time representation of a signal.

3.8 DISCUSSION

Figure 3.6 displays the periodogram for PPG and provides the frequency information of a processed signal. The example given in Figure 3.6 shows some low frequency noise. The spectrogram, on the other hand, provides

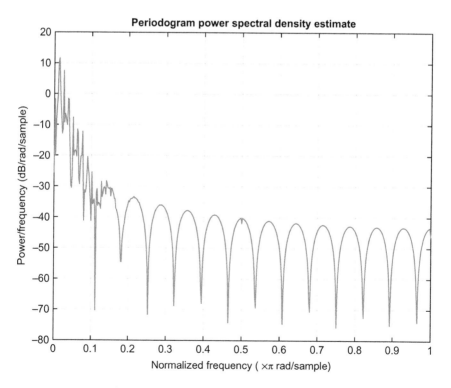

FIGURE 3.6 Periodogram for a PPG signal as shown in Figure 3.1.

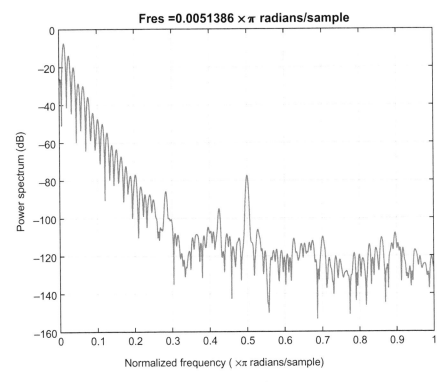

FIGURE 3.7   Spectrogram for a PPG signal as shown in Figure 3.1.

different frequency information about the signal, which is the frequency range that holds most of the energy of the processed signal. For example, most of the energy for the PPG signal is in the range 0–10 Hz, as shown in Figure 3.7. The MATLAB function that can be used to calculate the periodogram is:

```
1 >> periodogram(PPG);
```

The spectrogram of frequencies of PPG as they vary with time is shown in Figure 3.7. The MATLAB function that can be used to calculate the spectrum is:

```
1 >> pspectrum(PPG);
```

A continuous wavelet transform is used with the "Bump" and "Amor" wavelets, as shown in Figures 3.8 and 3.9, respectively. However, the "Bump" wavelet visual presentation did not provide any information

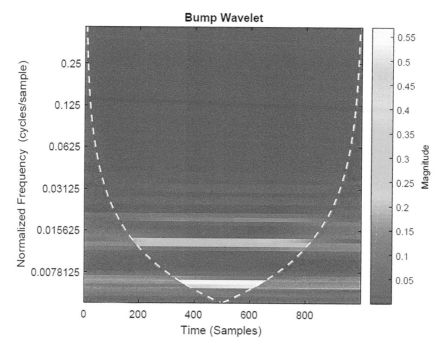

FIGURE 3.8   Bump wavelet for a PPG signal as shown in Figure 3.1.

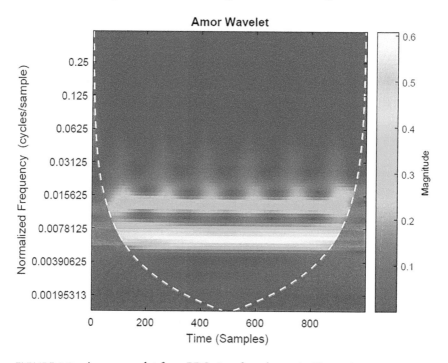

FIGURE 3.9   Amor wavelet for a PPG signal as shown in Figure 3.1.

about the signal, whereas the "Amor" wavelet localized the heartbeats. The MATLAB functions used to apply these steps are:

```
1  >> cwt(PPG, 'bump');
2  >> figure, cwt(PPG, 'amor');
```

The *eventogram* algorithm (cf. Figure 3.10) provides useful information about the PPG signals, information that was not provided by the standard methods discussed. For PPG signals, the window sizes (W_1 and W_2) were also varied, and the maximum power value was found at $W_1 = 41$ samples and $W_2 = 61$ samples. Standard visualization methods failed to provide insight into the events within the PPG signals.

The *eventogram* described uses two time–domain window sizes to describe biomedical signals, W_1 and W_2. The time–time representation provided by this method is different from the time–frequency representation provided by the spectrogram and wavelet methods. This time–time representation will help clinicians to understand the dominant durations, and to visualize related morphologies. The *eventogram* is more informative and, consequently, provides more insight into biomedical signal analysis when compared to the existing standard methods.

The application of the *eventogram* methodology on PPG signals is demonstrated in Figure 3.10. Dominant events are marked by red pixels where the combination of W_1 and W_2 scored the highest power. Once W_1 and W_2 are identified for the most dominant event, the next step is to visualize its corresponding waveform in terms of duration and morphology, as shown in Figure 3.11. The *eventogram* detected the systolic wave as the most dominant event in the PPG signal. Note that examination of the second highest power in the *eventogram*, the third highest, and so on, also makes it possible to check their corresponding dominant events.

The *eventogram* is also advantageous in that the search area can be customized or automated. For example, the search ranges for the processed PPG signal were 1–65 samples for W_1 and W_2. In Figure 3.12, the search areas are automatically adjusted to equal the sampling frequency, and the step sizes are automatically adjusted to be the *log* of the sampling frequency—the value will then be rounded toward negative infinity. Note, the window lengths for the processed PPG signal (cf. Figure 3.12) are 367 samples with a step size of five samples for PPG data length of 20 s.

Analysis of the regular heart rhythm biosignal with the *eventogram* is simple and efficient. The regularity of the heartbeats (beats are repeated in an equally spaced pattern) helps the time–domain threshold

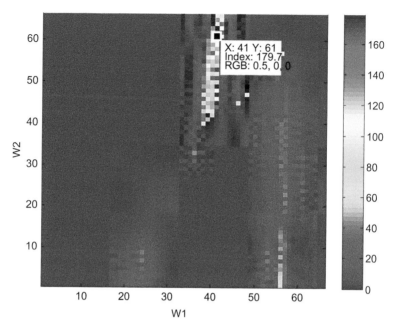

FIGURE 3.10 Eventogram for a PPG signal modified from Elgendi (2016).[58] *Notes*: Use of the color **red** indicates a strong dominance of a main event; use of the color **blue** indicates the non-existence of a main event. The *eventogram* is stored as a two-dimensional (W_1-by-W_2) array of integers in the range [1, length(colormap)]; the colormap is a W_1-by-3 matrix of real numbers between 0.0 and 1.0. The *eventogram* returns the index value in terms of W_1 and W_2 for the most dominant event that is associated with the red color or **RGB** (0.5,0,0)—the **RGB** color model is an additive color model in which red, green, and blue light are added together in various ways to reproduce a broad array of colors.

methodologies to detect main events successfully. The normal sinus rhythm[59] (regular heart rhythm) has a constant rhythm and the occurrence of the next beat is predictable. Systolic peaks are easily detected in a regular heart rhythm in PPG signals using the eventogram as shown in Figure 3.12a. Moreover, the *eventogram* detects main events in biomedical signals with irregular rhythms, as shown in Figure 3.12c. The *eventogram* is thus suitable for any quasi-periodic biomedical signal due to its robustness in the face of irregular rhythm and non-stationary effects (cf. Figure 3.12b–d), and low signal-to-noise ratio (cf. Figure 3.12d).

Precise information on event morphology and duration is not provided by the current standard visualization methods. Such methods rely on qualitative examination alone. The eventogram, on the other hand, relies on both qualitative and quantitative analysis. This new method uses

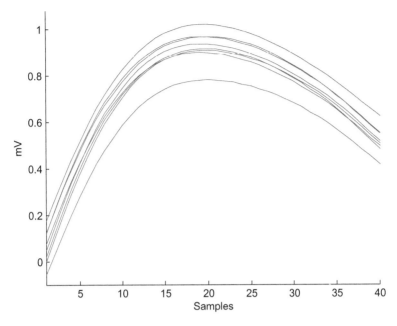

FIGURE 3.11 The main events (superimposed) detected by the *eventogram*s as shown in Figure 3.10. This illustration is based on the highest value of power scored with W_1 and W_2. This plot is modified from Elgendi (2016).[58]

different window size choices, and dominant events can be quantitatively and qualitatively visualized. This provides more precise insight into biomedical signal characteristics. To validate the concept, application, and performance of the *eventogram*, rigorous testing over multiple data sets for detecting different events were conducted as follows:

- For systolic wave detection in PPG signals: The *eventogram*-based systolic wave detector has two additional steps to the *eventogram*: one step at the beginning before applying the *eventogram* (filtering) and one at the end after applying the *eventogram* (thresholding). The *eventogram* is considered to be the feature extraction step. The *eventogram*-based systolic wave detection algorithm, with $W_1 = 111$ ms and $W_2 = 667$ ms, was evaluated using 40 records after three heat stress simulations, containing 5,071 heartbeats, with an overall SE of 99.89% and the +P was 99.84%.[53]

- For *a* and *b* wave detection in PPG signals: The *eventogram*-based *a* and *b* waves detector has two additional steps to the *eventogram*:

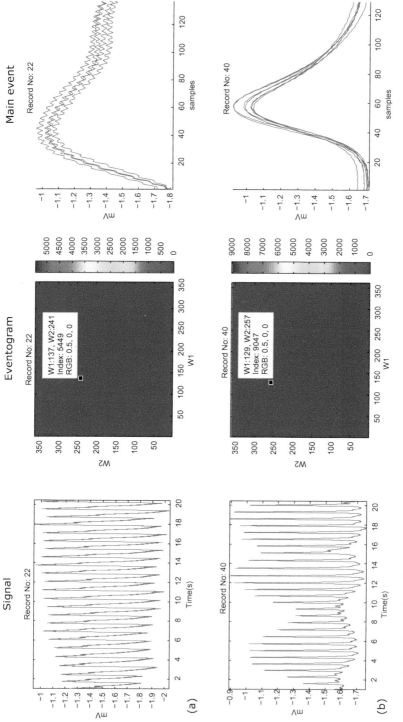

FIGURE 3.12 Examples of *eventogram* output using PPG signals modified from Elgendi (2016).[58] *Notes*: (a) PPG waveforms with a salient dicrotic notch; (b) PPG waveforms with a nonsalient dicrotic notch.

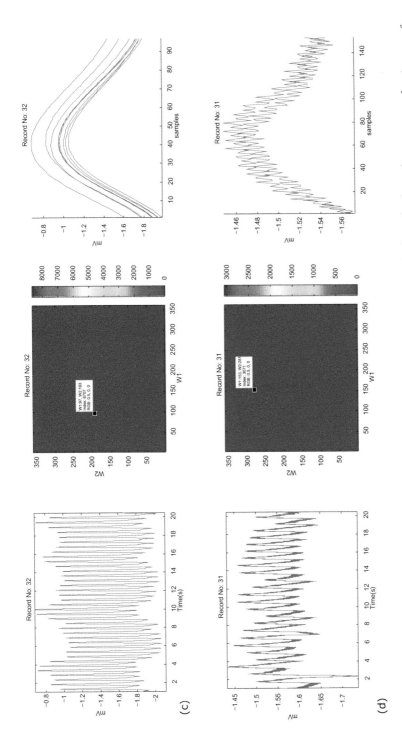

FIGURE 3.12 (c) a PPG signal measured after simulated heat stress; and (d) a noisy PPG signal. The **red** color indicates strong dominance of a main event while the **blue** color indicates the non-existence of a main event. The *eventogram* is stored as a two-dimensional (W_1-by-W_2) array of integers in the range [1, length(colormap)]; a colormap is a W_1-by-3 matrix of real numbers between 0.0 and 1.0. The *eventogram* returns the index value in terms of W_1 and W_2 for the most dominant event that is associated with the **red** color or **RGB** (0.5,0,0)—the **RGB** color model is an additive color model in which red, green, and blue light are added together in various ways to reproduce a broad array of colors.

one step at the beginning before applying the *eventogram* (filtering) and one at the end after applying the *eventogram* (thresholding). The *eventogram* is considered to be the feature extraction step. The *eventogram*-based *a* wave detection algorithm, with W_1 = 175 ms and W_2 = 1000 ms, demonstrated overall SE of 99.78%, +P of 100% over signals that suffer from: (i) non-stationary effects; (ii) irregular heartbeats; and (iii) low amplitude waves. In addition, the *b* detection algorithm (based on the detection of *a* waves) achieved an overall SE of 99.78% and a +P of 99.95%.[51]

- For *c*, *d*, and *e* wave detection in PPG signals: The *eventogram*-based *c*, *d*, and *e* waves detector has three additional steps to the *eventogram*: two steps at the beginning before applying the *eventogram* (filtering and removal of the *ab* segment) and one at the end after applying the *eventogram* (thresholding). The *eventogram* is considered to be the feature extraction step. The performance of the *eventogram*-based *c*, *d*, and *e* waves detector, with W_1 = 5 ms and W_2 = 15 ms, was tested on 27 PPG records collected during rest and after two hours of exercise, resulting in 97.39% SE and 99.82% +P.[52]

Finally, *eventogram*-based detectors succeeded in detecting systolic and *a*, *b*, *c*, *d*, *e* waves in PPG signals. Results show that the *eventogram*-based detectors are promising in terms of computational complexity and efficiency. Elgendi et. al presented the eventogram as a proof-of-concept work and it may fail when applied to other signals; thus, the eventogram needs to be used in the scientific community for performance evaluation. Temporal and spatial visualizations are used with the eventogram and this combination is a significant component of this method. Advantages of the eventogram include its customizability. Utilizing two moving averages yields efficient temporal filtering to the processed biomedical signal where it is statistically stationary at least over the duration of the moving window.

The *eventogram* was tested over one-dimensional biomedical signals that were collected and analyzed independently, covering temporal and spatial visualization. Multiple signals collected simultaneously need to be tested to determine the vectorial and interactive visualizations capability of the *eventogram*. Applying this method to different signal types, such as the optical signal and the communication signal, will improve generalization across disciplines.

3.9 SUMMARY

Analyzing a biomedical signal waveform in more detail is preceded by event detection and analysis. The *eventogram* uses a two event-related moving average approach and provides important information, such as the duration of the most dominant event and its morphology. New information and insight into the processed signal is also provided, given these features are not provided by current standards. Dominant events identified by this method provide clearer information for researchers and clinicians that will consequently improve their decision-making.

CHAPTER 4

Pre-processing of PPG Signals

D ue to the varying environments in which PPG-based portable or wearable devices are used and, consequently, their varying suscepti-bility to noise interference, the effective processing of PPG signals is chal-lenging. Thus, the aim of this chapter is to give an overview on how to design and implement a digital filter, rather than a hardware filter, using MATLAB for PPG applications.

4.1 LEARNING OBJECTIVES

The learning objectives of this chapter are to:

- Learn about the different types of filter

- Develop an understanding of the frequency response of digital filters

- Gain insight into how to select a suitable filter

4.2 FILTER TYPES

Many filters can be developed to remove undesired information from raw PPG signals. However, here we will focus only on nine filters. Usually, the design and selection of filters would be based on time domain or frequency domain output. Optimally, both time and frequency domains need to be explored to provide a robust filter for a particular application. The time domain aspect is easy; after applying the filter to the time-series PPG signal, the output can readily be inspected visually. However, the frequency changes in the filtered PPG signal are difficult to track and recognize. Therefore, it is important to understand the behavior of a filter in the frequency domain by exploring the *frequency response*. The *frequency response* is a measure of magnitude and phase following the injection of a sine wave.

As shown in Figure 4.1, there are main areas to consider when evaluating a filter. In the case a low pass filter, these are: the pass-band, transition, and stop-band. Ideally, the pass-band and stop-band ripples need to be flat (in other words, no ripples whatsoever), and the transition between bands should be sharp (in other words, closer to be a perpendicular line).

The smooth fall from stop-band to stop-band is called *roll-off*. Also, if there are *no* oscillations in the pass-band or stop-band, then the band is called *monotonic*. There is always a tradeoff between the roll-off and monotonicity. However, how to select the optimal filter with a particular roll-off and monotonicity depends on the application. The time domain analysis will play a major role in justifying the filter selection.

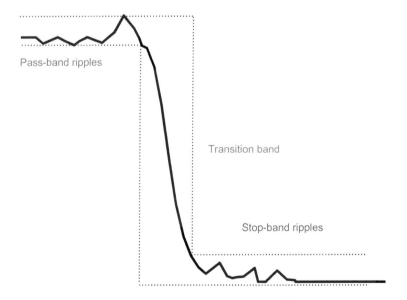

FIGURE 4.1   Bands of interest within the frequency response of a low pass filter.

4.2.1 Moving Average (MA) Filter

The moving average filter (MA) is the most common filter in the field of biomedical signal processing. It is easy to implement, easy to understand, and efficient in reducing random noise. The mathematical equation of an MA can be written as follows:

$$y[i] = 1/N \sum_{i=1}^{N} x_{i+j}, \tag{4.1}$$

where x is the raw PPG signal, y is the filtered PPG signal, and N is the number of points in the average. For example, an MA with a window length of four samples can be written as follows:

$$y[1] = 1/4 \big[x[1] + x[2] + x[3] + x[4] \big]. \tag{4.2}$$

The implementation of this step in MATLAB is easy and can be carried out as follows:

```
1  >> % Implementing the moving average using a simple
       for loop
2  >> windowSize = 4;
3  >> Raw_Sig = "your PPG signal";
4  >> figure, plot (Raw_Sig);
5  >> for i=1:length(x)-WindowSize
6         Filtered_Sig(i) = 1/WindowSize *(Raw_Sig(i)
              + Raw_Sig(i+1) + Raw_Sig(i+2) +
              Raw_Sig(i+3));
7  >> end
8  >> figure, plot (Filtered_Sig);
9  >> xlabel('Samples');
10 >> ylabel('Amplitude');
11 >> title('Moving Average')
```

It is worth noting that the MA can be calculated using *convolution* with a simple filter kernel. For example, the kernel of an MA with a window length of four samples, can be set as an array with the following values [0, 0, 1/4, 1/4, 1/4, 1/4, 0, 0]. The implementation of an MA with a window length of four samples using convolution can be carried in MATLAB as follows:

```
1  >> % Implementing the moving average using Convolution
2  >> windowSize = 4;
```

```
 3 >> Raw_Sig = "your PPG signal";
 4 >> figure, plot (Raw_Sig);
 5 >> kernel = [0,0, 1/4, 1/4, 1/4, 1/4, 0, 0];
 6 >> Filtered_Sig = conv(Sig, kernel, 'same');
 7 >> figure, plot (Filtered_Sig);
 8 >> xlabel('Samples');
 9 >> ylabel('Amplitude');
10 >> title('Moving Average')
```

As can be seen, the kernel of the convolution for developing an MA is a rectangular pulse having an area of 1. There are many approaches to developing an MA, and MATLAB has multiple functions that can be used, such as *smooth* and *movmean*. Note that the basic parameter for an MA is the window length of n samples.

In time domain analysis, MAs are useful and effective; however, this is not so with the frequency domain. Let us now see the frequency response of an MA:

```
 1
 2 >> % Implementing the frequency response of Moving
      Average filter
 3 >> % Sampling frequency in Hz
 4 >> Fs = 200;
 5 >> windowSize = 5;
 6 >> num = (1/windowSize)*ones(1,windowSize);
 7 >> dend = 1;
 8 >> % Logarithimc scale
 9 >> L=logspace(0,2);
10 >> % call the Freqency response function
11 >> Z=freqz(num,dend,L,Fs);
12 >> % Compute and display the magnitude response
13 >> figure, semilogx(L,abs(Z),'K');
14 >> grid;
15 >> xlabel('Hz');
16 >> ylabel('Gain');
17 >> title('Moving Average')
```

As can be seen in Figure 4.2, the roll-off is very slow and the stop-band ripples are high. We see a flat pass-band; however, the MA cannot separate one band of frequencies from another. Therefore, if the application is focusing on separating different bands with a high degree of accuracy,

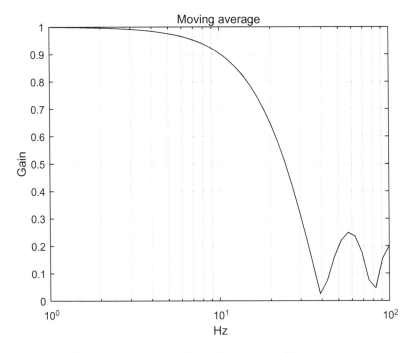

FIGURE 4.2 Frequency response of a moving average filter.

then MA alone is not useful. Perhaps other steps are needed before or after to provide an optimal frequency response.

The reader may conclude that, as the MA performance is poor in the frequency domain, then either MAs are not useful, or should not be used at all. This is not the case: MAs are exceptionally good smoothing filters. Again, it depends on the application and how the MAs are implemented in the main algorithm.

For noise removal, the moving average filter makes an excellent job of reducing the amount of noise within the PPG signal, as data samples are treated equally. From visual inspection, in the time domain, no filter performs better than the simple moving average. However, we cannot say moving average filters are optimal for all applications (e.g. noise removal, heart rate detection, and so on). The application dictates what is needed from the filtration process. Sometimes, aggressive filtering is required if we are simply looking for a trend, whereas soft filtering such as that performed by the moving filter can be enough to achieve the aim. Therefore, examining both time and frequency domains is essential for each application.

4.2.2 Butterworth Filter (Butter)

A butter filter of order *m*: The MATLAB function used to implement this step was *butter*. Figure 4.3 presents the behaviour of the Butterworth filter.

```
1
2 % Implementing the frequency response of Butterworth
   filter
3 % Sampling frequency in Hz
4 >> Fs=200;
5 >> Fc=6/(Fs/2);
6 >> m=6;
7 >> % Butterworth filter
8 >> [num dend]=butter(m,Fc);
9 >> % Logarithimc scale
10 >> L=logspace(0,2);
11 >> % call the Freqency response function
12 >> Z=freqz(num,dend,L,Fs);
13 >> % Compute and display the magnitude response
14 >> figure; semilogx(L,abs(Z),'K');
```

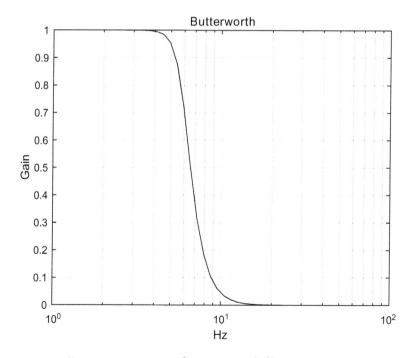

FIGURE 4.3   Frequency response of a Butterworth filter.

```
15 >> grid;
16 >> xlabel('Hz');
17 >> ylabel('Gain');
18 >> title('Butterworth')
```

4.2.3 Chebyshev Filter (Cheby I and Cheby II)

A Chebyshev I filter of order m: The MATLAB functions used to implement the two filters discussed in this section were *cheby1* and *cheby2*. The function *cheby1*(n, Rp, Wp) returns the transfer function coefficients of an nth-order low-pass digital Chebyshev I filter with normalized passband edge frequency R_c and F_c decibels of peak-to-peak passband ripple. Figure 4.4 presents the behaviour of the Chebyshev filters.

```
 1 >> % Implementing the frequency response of Cheby I
      filter
 2 >> % Sampling frequency in Hz
 3 >> Fs=200;
 4 >> Fc=6/(Fs/2);
 5 >> m=6;
 6 >> Rs=18;
 7 >> % cheby1 filter
 8 >> [num, dend]=cheby1(m,Rs,Fc);
 9 >> % Logarithimc scale
10 >> L=logspace(0,2);
11 >> % call the Freqency response function
12 >> Z=freqz(num,dend,L,Fs);
13 >> % Compute and display the magnitude response
14 >> figure, semilogx(L,abs(Z),'K');
15 >> grid;
16 >> xlabel('Hz');
17 >> ylabel('Gain');
18 >> title('Chebyshev 2')
```

```
 1 >> % Implementing the frequency response of Cheby
      II filter
 2 >> % Sampling frequency in Hz
 3 >> Fs=200;
 4 >> Fc=6/(Fs/2);
 5 >> m=6;
 6 >> Rs=18;
 7 >> % cheby2 filter
 8 >> [num, dend]=cheby2(m,Rs,Fc);
 9 >> % Logarithimc scale
10 >> L=logspace(0,2);
```

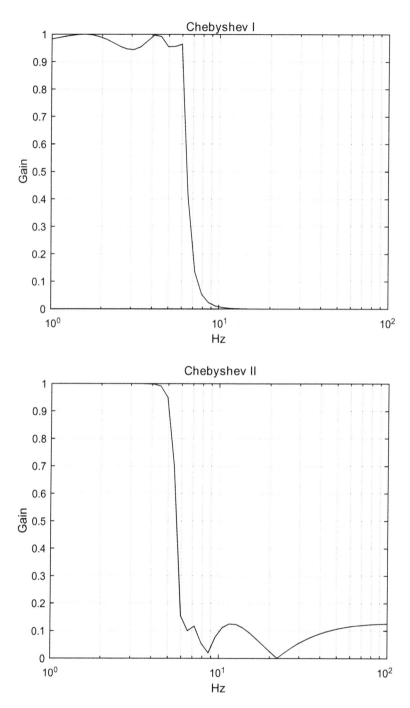

FIGURE 4.4   Frequency response of Chebyshev filters.

```
11 >> % call the Freqency response function
12 >> Z=freqz(num,dend,L,Fs);
13 >> % Compute and display the magnitude response
14 >> figure, semilogx(L,abs(Z),'K');
15 >> grid;
16 >> xlabel('Hz');
17 >> ylabel('Gain');
18 >> title('Chebyshev 2')
```

4.2.4 Elliptic Filter (Ellip)

An ellip filter of order m: The MATLAB function used to implement this step was *ellip*. Figure 4.5 presents the behaviour of the Elliptic filter.

```
1 >> % Implementing the frequency response of Elliptic
     filter
2 >> Fs=200;
3 >> Fc=6/(Fs/2);
4 >> m=6;
5 >> Rp=0.5;
6 >> Rc=20;
7 >> % Butterworth filter
8 >> [num, dend]=ellip(m,Rp,Rc,Fc);
```

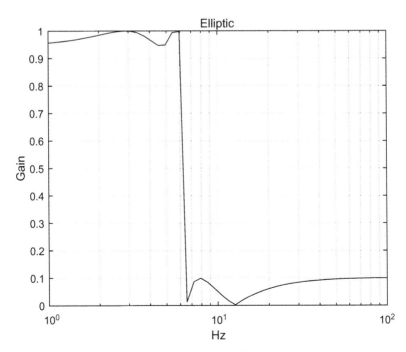

FIGURE 4.5   Frequency response of an elliptic filter.

```
 9 >> % Logarithimc scale
10 >> L=logspace(0,2);
11 >> % call the Freqency response function
12 >> Z=freqz(num,dend,L,Fs);
13 >> % Compute and display the magnitude response
14 >> figure, semilogx(L,abs(Z),'K');
15 >> grid;
16 >> xlabel('Hz');
17 >> ylabel('Gain');
18 >> title('Elliptic')
```

4.2.5 General Comment

From the time-domain perspective, with regard to the assessment of signal quality based on time-domain morphology, a recent study[60] observed that the Chebyshev II filter improves the PPG signal quality more effectively than other filters, and therefore we only made a comparison between the Butterworth filter (the gold standard filter) and the Chebyshev II filter, as shown in Figure 4.6.

The Chebyshev II is the optimal filter due to its superior frequency selectivity and flat passband. From the example shown in Figure 4.6, the Chebyshev II improved most of the the morphology of the PPG waveforms. The left plot in Figure 4.6 shows the impulse response difference between the Butterworth (red line) and Chebyshev II (black line) filters. The right plot in Figure 4.6 shows the Butterworth bandpass filtered (red

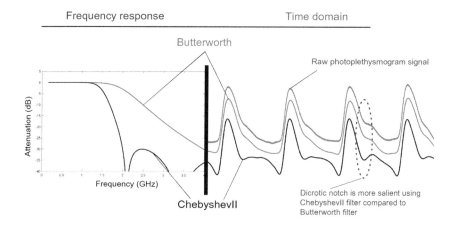

FIGURE 4.6 Filter impact on PPG morphology. This figure is adopted from Elgendi et al. (2019).[46]

line) and the Chebyshev II bandpass (black line) filtered PPG signals of the raw PPG signal (blue line). It is clear that the Chebyshev II filter is better able to emphasize the difference between the systolic and diastolic waves, compared to the Butterworth filter.

It is worth noting that there are different types of filters that can be implemented in MATLAB, such as:

- Median filter (MF): An MF filter with a window length of n samples. The MATLAB function used to implement this step was *medfilt1*.

- Finite impulse response filter (FIR-hamming and FIR-ls): An FIR filter of order m. The MATLAB functions used to implement these two filters were *fir1* and *designfilt*, respectively.

- Wavelet de-noising filter (Wavelet): The wavelet transform is a powerful tool for signal and image processing, and has been successfully applied to many scientific fields, such as signal processing, image compression, computer graphics, and pattern recognition. Wavelet de-noising consists of three steps: wavelet decomposition, the thresholding of detail coefficients, and reconstruction. The MATLAB function used to implement this step was *wden*.

4.3 FILTER DESIGN

4.3.1 Low-Pass Filter

The low-pass filter is a filter that passes PPG signals of a frequency lower than a selected cutoff frequency and reduces (ideally, eliminates) frequencies higher than the cutoff frequency. Below is an implementation of different low-pass filters in MATLAB.

```
1
2  %% Parameters:
3  % filter_type ———————filter type
4  % order————————————filter order (level –
      Wavelet or point——Fir)
5  % raw_data—————————raw PPG signal
6  % Fs———————————————sample frequency
7  % fc———————————————Cutoff Frequency
8
9  function [filtered_data] = PPG_Lowpass(raw_data,
      filter_type,order,Fs,fc)
10 Fn = Fs/2;
11
```

```
12  switch filter_type
13      case 1
14          [A,B,C,D] = butter(order,fc/Fn,'low');
15          [filter_SOS,g] = ss2sos(A,B,C,D);
16          filtered_data = filtfilt(filter_SOS,g,
            raw_data);
17      case 2
18          [A,B,C,D] = cheby1(order,0.1,fc/Fn,'low');
19          [filter_SOS,g] = ss2sos(A,B,C,D);
20          filtered_data = filtfilt(filter_SOS,g,
            raw_data);
21      case 3
22          [A,B,C,D] = cheby2(order,20,fc/Fn,'low');
23          [filter_SOS,g] = ss2sos(A,B,C,D);
24          filtered_data = filtfilt(filter_SOS,g,
            raw_data);
25      case 4
26          [A,B,C,D] = ellip(order,0.1,30,fc/Fn,'low');
27          [filter_SOS,g] = ss2sos(A,B,C,D);
28          filtered_data = filtfilt(filter_SOS,g,
            raw_data);
29      case 5
30          d = fir1(order,fc/Fn,'low');
31          filtered_data = filtfilt(d,1,raw_data);
32          filter_SOS = d;
33      case 6
34          d = designfilt('lowpassfir','FilterOrder',
                order,' PassbandFrequency',fc,'Stopband
                Frequency',fc+0.2,'DesignMethod','ls',
                'SampleRate',sample_freq);
35          filtered_data = filtfilt(d,raw_data);
36
37      case 7
38          filtered_data = smooth(raw_data,order);
39
40      case 8
41          filtered_data = medfilt1(raw_data,order);
42
43      case 9
44          filtered_data= wden(raw_data,'modwtsqtwolog',
                's','mln',order,'db2'); %Wavelet level:
                order
45
46  end
47
48  end
```

4.3.2 High-Pass Filter

The high-pass filter is a filter that passes PPG signals of a frequency higher than a selected cutoff frequency and reduces (ideally, eliminates) frequencies lower than the cutoff frequency. Below is an implementation of different high-pass filters in MATLAB.

```
1
2  %% Parameters:
3  % filter_type ──────────────filter type
4  % order──────────────────────filter order (level -
     Wavelet or point──Fir)
5  % raw_data────────────────────raw PPG signal
6  % Fs──────────────────────────sample frequency
7  % fc──────────────────────────Cutoff Frequency
8
9
10 function [filtered_data,filter_SOS] = PPG_Highpass
       (raw_data,filter_type,order,sample_freq,fc)
11 Fn = sample_freq/2;
12
13 switch filter_type
14     case 1
15         [A,B,C,D] = butter(order,fc/Fn,'high');
16         [filter_SOS,g] = ss2sos(A,B,C,D);
17         filtered_data = filtfilt(filter_SOS,g,
           raw_data);
18     case 2
19         [A,B,C,D] = cheby1(order,0.1,fc/Fn,'high');
20         [filter_SOS,g] = ss2sos(A,B,C,D);
21         filtered_data = filtfilt(filter_SOS,g,
           raw_data);
22     case 3
23         [A,B,C,D] = cheby2(order,20,fc/Fn,'high');
24         [filter_SOS,g] = ss2sos(A,B,C,D);
25         filtered_data = filtfilt(filter_SOS,g,
           raw_data);
26     case 4
27         [A,B,C,D] = ellip(order,0.1,30,fc/Fn,'high');
28         [filter_SOS,g] = ss2sos(A,B,C,D);
29         filtered_data = filtfilt(filter_SOS,g,
           raw_data);
30     case 5
31         d = fir1(order,fc/Fn,'high');
32         filtered_data = filtfilt(d,1,raw_data);
```

```
33      case 6
34          d = designfilt('highpassfir','FilterOrder',
                order,'StopbandFrequency',fc-0.2,
                'PassbandFrequency',fc,'DesignMethod',
                'ls','SampleRate',sample_freq);
35          filtered_data = filtfilt(d,raw_data);
36      case 7
37          filtered_data = smooth(raw_data,order);
38      case 8
39          filtered_data = medfilt1(raw_data,order);
40      case 9
41          filtered_data= wden(raw_data,'modwtsqtwolog',
                's','mln',order,'db2'); %Wavelet level :
                order
42 end
43
44 end
```

4.3.3 Band-Pass Filter

The bandpass filter is a filter that passes PPG signals within two selected cutoff frequencies and reduces (ideally, eliminates) frequencies that are not within the range of the cutoff frequencies. Below is an implementation of different bandpass filters in MATLAB.

```
1
2 %% Parameters :
3 % filter_type ------------filter type
4 % order------------------filter order
5 % raw_data---------------raw PPG signal
6 % Fs--------------------sampling frequency
7 % fL--------------------lower cutoff frequency
8 % fH--------------------higher cutoff frequency
9
10 function [filtered_data] = PPG_Bandpass(raw_data,
       filter_type,order,Fs,fL,fH)
11
12 Fn = Fs/2;
13
14 switch filter_type
15     case 1
16         [A,B,C,D] = butter(order,[fL fH]/Fn);
17         [filter_SOS,g] = ss2sos(A,B,C,D);
18         filtered_data = filtfilt(filter_SOS,g,
            raw_data);
```

```
19      case 2
20          [A,B,C,D] = cheby1(order,0.1,[fL fH]/Fn);
21          [filter_SOS,g] = ss2sos(A,B,C,D);
22          filtered_data = filtfilt(filter_SOS,g,
            raw_data);
23      case 3
24          [A,B,C,D] = cheby2(order,20,[fL fH]/Fn);
25          [filter_SOS,g] = ss2sos(A,B,C,D);
26          filtered_data = filtfilt(filter_SOS,g,
            raw_data);
27      case 4
28          [A,B,C,D] = ellip(order,0.1,30,[fL fH]/Fn);
29          [filter_SOS,g] = ss2sos(A,B,C,D);
30          filtered_data = filtfilt(filter_SOS,g,
            raw_data);
31      case 5
32          d = fir1(order,[fL fH]/Fn,'bandpass');
33          filtered_data = filtfilt(d,1,raw_data);
34      case 6
35          d = designfilt('bandpassfir','FilterOrder',
              order,'StopbandFrequency1',fL-0.2,
              'PassbandFrequency1',fL, ...
              'PassbandFrequency2',fH,'Stopband
              Frequency2',fH+2,'DesignMethod',
              'ls','SampleRate',sample_freq);
36
37          filtered_data = filtfilt(d,raw_data);
38      case 7
39          filtered_data = smooth(raw_data,order);
40      case 8
41          filtered_data = medfilt1(raw_data,order);
42      case 9
43          filtered_data= wden(raw_data,'modwtsqtwolog',
            's','mln', order,'db2'); %Wavelet level:
            order
44  end
45
46  end
```

4.3.4 Band-Stop Filter

The function of the Band-stop filter is the opposite of the bandpass filter, as it does not pass PPG signals within two selected cutoff frequencies but, instead, passes frequencies that are not within the range of the cutoff frequencies. Sometimes, a band-stop filter has a narrow stopband called a *notch filter*. Below is an implementation of different band-stop filters in MATLAB.

```
%% Parameters:
% filter_type ——————————filter type
% order——————————————————filter order
% raw_data————————————————raw PPG signal
% Fs——————————————————————sampling frequency
% fL——————————————————————lower cutoff frequency
% fH——————————————————————higher cutoff frequency

function [filtered_data] = PPG_Bandstop(raw_data,
    filter_type, order,Fs,fL,fH)

Fn = Fs/2;

switch filter_type
    case 1
        [A,B,C,D] = butter(order,[fL fH]/Fn,
        'stop');
        [filter_SOS,g] = ss2sos(A,B,C,D);
        filtered_data = filtfilt(filter_SOS,g,
        raw_data);
    case 2
        [A,B,C,D] = cheby1(order,0.1,[fL fH]/Fn,
        'stop');
        [filter_SOS,g] = ss2sos(A,B,C,D);
        filtered_data = filtfilt(filter_SOS,g,
        raw_data);
    case 3
        [A,B,C,D] = cheby2(order,20,[fL fH]/Fn,
        'stop');
        [filter_SOS,g] = ss2sos(A,B,C,D);
        filtered_data = filtfilt(filter_SOS,g,
        raw_data);
    case 4
        [A,B,C,D] = ellip(order,0.1,30,[fL fH]/Fn,
        'stop');
        [filter_SOS,g] = ss2sos(A,B,C,D);
        filtered_data = filtfilt(filter_SOS,g,
        raw_data);
    case 5
        d = fir1(order,[fL fH]/Fn,'stop');
```

```
33          filtered_data = filtfilt(d,1,raw_data);
34      case 6
35  end
36
37  end
```

4.4 CONVOLUTION

Convolution is a mathematical operation on two functions (*f* and *g*), the result expressing how the shape of one is modified by the other. Convolution can be used as a filter that improves the quality of either the whole PPG signal, or one PPG waveform.

4.4.1 Improving PPG Beat Quality

The following MATLAB code shows how to improve the quality of one PPG waveform using convolution. Figure 4.7 shows the PPG waveform template (*f*), the PPG beat that needs improvement (*g*), and the result of the convolution between them.

```
1   f = template;      % A template that is considered
                of excellent quality
2   g = beat;          % One PPG beat that needs to be
                            improved
3         g = resample(beat,length(f),length(g));
                %resampling the signal to the same
                length as the template
4         conv_f_g = conv(f, g);
5         conv_g_f = conv(g, f);
6
7         figure;
8         subplot(4,1,1)
9         plot(f,'k-');
10        title('f: Template');
11        subplot(4,1,2);
12        plot(g,'r-');
13        title('g: Raw signal');
14        subplot(4,1,3);
15        plot(conv_f_g);
16        title('f * g');
17        subplot(4,1,4);
18        plot(conv_g_f);
19        title('g * f');
20        suptitle('Convolution');
```

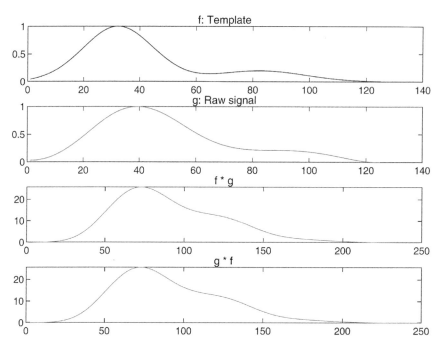

FIGURE 4.7 Convolution between a PPG template and a PPG waveform. Here, *f* refers to the PPG waveform template where *g* refers to the PPG beat that needs improvement.

The different waveforms of template will achieve varying results. The length of the signal and template also affect the result, and it is better they are the same length. The function *resample* is used to resample the signal to the same length as the template.

4.4.2 Filtering PPG Signal

The following MATLAB code shows how to filter the PPG signal using convolution between a PPG waveform template (*f*) and a raw PPG signal (*g*, a PPG signal that needs to be filtered). Figure 4.8 shows the PPG waveform template, the signal that needs to emphasize certain events (based on the application), and the result of the convolution between them.

```
1    f = template;      % A PPG waveform template
2    g = signal;        % A raw PPG signal
3    conv_f_g = conv(f, g); % same result as conv_g_f =
         conv (g, f);
```

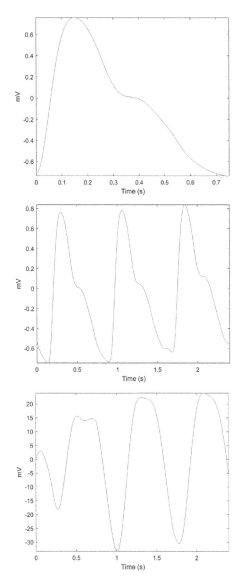

FIGURE 4.8 Convolution between a PPG template and raw PPG signal. The top plot is the PPG waveform template, the middle plot is the signal that needs improvement, the bottom figure is the result of their convolution.

It is clear that convolution can help to emphasize beats and make them more salient; for example, to detect a heart rate more accurately. Note that the template length does need to be the same length as the signal. In Section 4.4.1, there was a need to resample the template or the beat that needs improvement. This step is not required when using convolution.

4.5 CROSS-CORRELATION

Cross-correlation is a measure of similarity of two series as a function of the displacement of one relative to the other. The cross-correlation is similar to the convolution of two functions and could be used as a filter.

4.5.1 Filtering One PPG Beat

The following MATLAB code shows how to filter one PPG waveform using cross-correlation. Figure 4.9 shows the PPG waveform template (f), the PPG beat that needs improvement for a particular application (g), and the result of the cross-correlation between them.

The following code shows how to calculate the cross-correlation:

```
1 f = template;   % Template is an excellent PPG
      waveform
2 g = signal;   %Signal is the ppg beat that needs to
      be processed
```

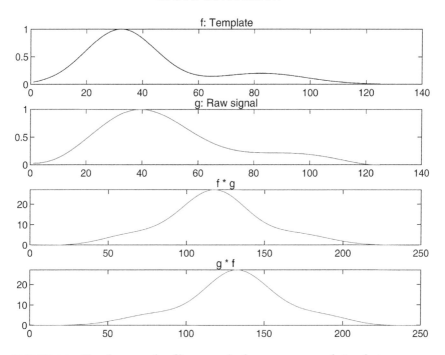

FIGURE 4.9   Simple example of how to calculate a cross-correlation between one beat signal and a template.

```
3 g = resample(signal,length(f),length(g));
      %resampling the signal to the same length as the
      template
4 xcorr_f_g = xcorr(f,g);
5 xcorr_g_f = xcorr(g,f);
6
7 figure;
8 subplot(4,1,1)
9 plot(f,'k-');
10 title('f: Template');
11 subplot(4,1,2);
12 plot(g,'r-');
13 title('g: Raw signal');
14 subplot(4,1,3);
15 plot(xcorr_f_g);
16 title('f * g');
17 subplot(4,1,4);
18 plot(xcorr_g_f);
19 title('g * f');
20 suptitle('Cross correlation');
```

Notes that the result of $f * g$ is different from $g * f$.

4.5.2 Filtering PPG Signal Quality

The following MATLAB code shows how to filter the PPG signal using cross-correlation between a PPG waveform template (f) and a raw PPG signal (g, a PPG signal that needs to be filtered). Figure 4.10 shows the PPG waveform template, the signal that needs improvement (based on the application), and the result of the cross-correlation between them.

```
1 f = template;      % A PPG waveform template
2 g = signal;        % A raw PPG signal
3 xcorr_f_g = xcorr(f, g);
4 xcorr_g_h = xcorr(g, f);
```

We can see that cross-correlation is different from convolution; the output of cross-correlation needs to be trimmed by half in terms of length, while the output of convolution does not need any trimming. However, both methods could be used to emphasize heartbeats to detect a heart rate more accurately. There is no rule of thumb here; both methods could be used,

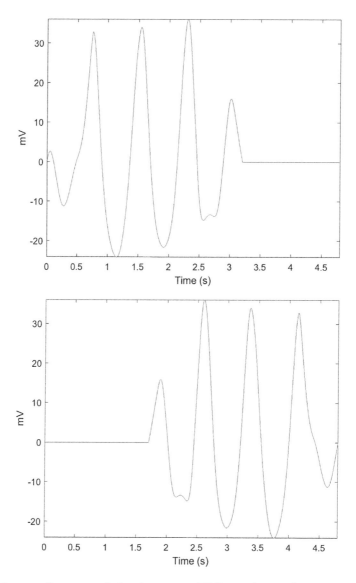

FIGURE 4.10 Cross-correlation between a PPG template and raw PPG signal. The top plot is the PPG waveform template, the middle plot is the signal that needs improvement, the bottom figure is the result of their cross-correlation.

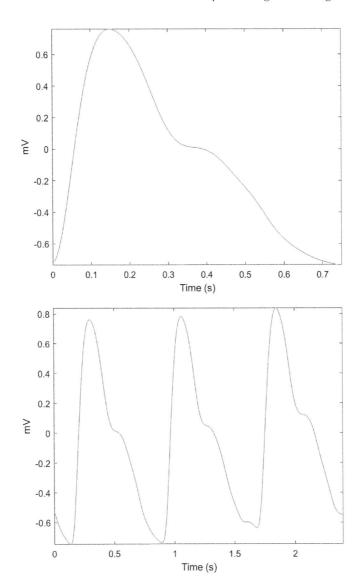

FIGURE 4.10   *Contined*

testing first to check performance based on the collected data, and the required application.

4.6 SUMMARY

There is a tradeoff between the flatness of the stop and pass bands, and the steepness of the transition. The *Butterworth* filter is commonly used as it has flat bands, but slow transition. The *elliptic* filter has the fastest transition, but has ripples in the pass and stop bands. This does not mean that the *elliptic* filter is the optimal choice for all PPG applications. Again, it depends on the application and whether the concern is more about trend, amplitude, or morphology. For example, results show that Chebyshev II can emphasize diastolic waves in PPG signals and is able to improve the signal quality, compared to the Butterworth filter. Other pre-processing techniques—such as convolution, cross-correlation, and auto-correlation—could be used depending on the application and what needs to be achieved.

CHAPTER 5

Signal Quality Assessment

Wearable devices are increasingly incorporating PPG sensors, and this means that PPG signals will be collected in a wide range of circumstances (walking, running, and so on). The collection of PPG signals in varying conditions will incur motion and noise artifacts that make analysis more difficult. This chapter discuss the challenges of PPG data collection and possible techniques to reduce extraneous noise in PPG signals.

5.1 LEARNING OBJECTIVES

The learning objectives of this chapter are to:

- Learn about the intrinsic weakness of using PPG for diagnosis

- Develop an understanding of the challenges around PPG signal quality assessment

- Gain insight into the practical implementation of signal quality indices and their reliability for PPG waveforms assessment

5.2 INTRODUCTION

The pulse oximeter is the most commonly used mobile monitoring device for measuring a patient's oxygen saturation levels and heart rate (HR).[61,62] Its popularity is due to its advantages as a non-invasive, inexpensive, and convenient screening tool that is remarkably easy to use and comfortable for patients. Although this tool is traditionally used to collect oxygen saturation measurements, the PPG signal collected using the pulse oximeter provides other important information through its signal waveform morphology.[63] For this reason, researchers are striving to maximize the utility of the PPG waveform characteristics to develop clinically useful devices.[64]

Recently, there has been growing interest in the real-time, wearable, and ambulatory monitoring of vital signs using pulse oximeter sensors. However, motion and noise artifacts are a serious obstacle in collecting clear signals to use for the clinical diagnosis of certain diseases and related ailments. Artifacts have been recognized as an intrinsic weakness of using PPG for diagnosis, as the noise can limit the practical implementation and reliability of real-time monitoring applications. Artifacts are the most common cause of false alarms, signal loss, and inaccurate measurements and diagonses.[65]

Although the clinical significance of PPG measurements has been well-investigated,[12,66–69], there is still a lack of studies focused on determining the optimal signal quality index (SQI) for assessing PPG signals, especially for mobile health applications. Several pulse oximetery manufacturers, such as Philips (Amsterdam, Netherlands), Nellcor-Medtronic (Dublin, Ireland), and Masimo (Irvine, California, USA), use the perfusion index as the gold standard for PPG signal quality assessment[70–73]. Recently, three SQIs have been tested for PPG quality assessment, including skewness,[74] kurtosis,[74,75] and Shannon entropy.[75] Other SQIs have also been shown to be useful for detecting artifacts in ECG signals.[76] However, no detailed quantitative results have been reported to verify the accuracy and suitability of SQIs for the successful detection of artifacts in PPG waveforms. In total, eight SQIs were investigated and comparison was made of the performance of the perfusion index to seven other SQIs. Moreover, the optimal SQI has been reported for assessing PPG signals.

5.3 ANNOTATION

Two independent annotators annotated the signals (106 PPG recordings, 60 s each) based on three groups: Group 1 (G1) corresponds to "excellent" for diagnosis, Group 2 (G2) corresponds to "acceptable" for diagnosis, and Group 3 (G3) corresponds to "unfit" for diagnosis. The annotation process was carried out over each 60 s PPG recording. Each annotator annotated the PPG signal based on the most dominant beat morphology quality within the signal. Since it is expected that each 60 s recording will have approximately 60 beats, a 60 s recording with 30 beats or more will be considered dominant within its designated group. For consistency, these groups were clearly predefined (an example is shown in Figure 5.1) for the annotators, who then adhered to the categories during the annotation process:

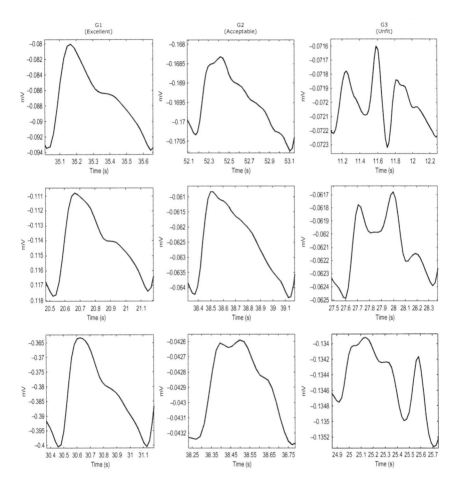

FIGURE 5.1   Annotation of photoplethysmogram (PPG) signals. This figure is adopted from Elgendi (2016).[84] *Notes*: Annota-ting the whole signal is based on the most dominant beat wave quality within the signal. The waves of the most dominant beats were categorized into three categories: G1 contains clear systolic beats and diastolic waveforms with dicrotic notches; G2 contains unclear systolic beats and diastolic waveforms without dicrotic notches; and G3 contains noisy waveforms. Each column represents a different group, with three examples showing the whole PPG signal (the lefthand side of each example) and its most dominant beat wave (the righthand side in each example).

1. *Excellent for diagnosis*: The "excellent" for diagnosis group (G1) includes only PPG signals where the systolic and diastolic waves are salient. If there are more than two waves, it is an indication of a noisy waveform, and therefore it may qualify for one of the next two types.

2. *Acceptable for diagnosis*: The "acceptable" for diagnosis group (G2) includes only PPG signals where the systolic and diastolic waves are *not* salient but where the HR can be determined.

3. *Unfit for diagnosis*: The "unfit" for diagnosis group (G3) includes only noisy PPG signals where the HR cannot be determined, and the systolic and diastolic waves cannot be distinguished.

The 106 PPG recordings were duplicated in the dataset to enable agreement between the annotators. Cohen's kappa coefficient (k) was used to measure agreement between annotators,[77] which is defined as:

$$k = \left(P_r(a) - P_r(e) \right) / \left(1 - P_r(e) \right) \tag{5.1}$$

where Pr(a) is the relatively observed agreement among annotators and Pr(e) is the hypothetical probability of chance agreement. An adjudication of discrepancies was carried out by an expert with more than a decade of experience in examining and processing PPGs to generate one annotation file for all PPG signals to be used in the training and classification stages. Here is the MATLAB implementation of the kappa coefficient:

```
1  >> n=sum(x(:)); %Sum of Matrix elements
2  >> x=x./n; %proportion
3  >> r=sum(x,2); %rows sum
4  >> s=sum(x); %columns sum
5  >> Ex=r*s; %expected proportion for random agree
6  >> pom=sum(min([r';s])); %maximum proportion
       observable
7  >> po=sum(sum(x.*f)); %proportion observed
8  >> pe=sum(sum(Ex.*f)); %proportion expected
9  >> k=(po-pe)/(1-pe); %Cohen's kappa
```

5.4 SIGNAL QUALITY INDICES

Eight Signal Quality Indices (SQIs) are commonly used in the field of signal processing, all of which have been rigorously tested and evaluated. Definitions of these SQIs (i.e. mathematical representations) and discussion of the implementation of each is essential to understanding their strengths and limitations, and to understanding when each is appropriate for use.

5.4.1 Perfusion (P_{SQI})

The perfusion index is the gold standard for assessing PPG signal quality.[70–73] It is the ratio of the pulsatile blood flow to the nonpulsatile or static blood in peripheral tissue. In other words, it is the difference in the amount of light absorbed through the pulse when light is transmitted through the finger, and is defined as follows:

$$P_{SQI} = \left[\left(y_{max} - y_{min} \right) / ||\bar{x}|| \right] \times 100 \tag{5.2}$$

where P_{SQI} is the perfusion index, $\bar{x}$ is the statistical mean of the x signal (raw PPG signal), and y is the filtered PPG signal.

```
1  P_SQI = (max(filtered_signal) - min(filtered_
      signal))/mean(signal) * 100;
```

5.4.2 Skewness (S_{SQI}):

Krishnan et al.[74] found that the skewness index is associated with corrupted PPG signals. Skewness is a measure of the symmetry (or the lack of it) of a probability distribution, which is defined as:

$$S_{SQI} = 1/N \sum_{i=1}^{N} \left[x_i - \hat{\mu}_x / \sigma \right]^3 \tag{5.3}$$

where $\hat{\mu}_x$ and σ are the empirical estimate of the mean and standard deviation of x_i, respectively, and N is the number of samples in the PPG signal.

```
1  S_SQI = skewness(signal);
```

5.4.3 Kurtosis (K_{SQI})

Recently, Selvaraj et al.[75] found that kurtosis is a good indicator for PPG signal quality. Kurtosis is a statistical measure used to describe the distribution of observed data around the mean. It represents a heavy tail and peakedness, or a light tail and a flatness of a distribution relative to the normal distribution, which is defined as:

$$K_{SQI} = 1/N \sum_{i=1}^{N} \left[x_i - \hat{\mu}_x / \sigma \right]^4 \tag{5.4}$$

where $\hat{\mu}_x$ and σ are the empirical estimate of the mean and standard deviation of x_i, respectively; and N is the number of samples in the PPG signal. A MATLAB function can be used as follows:

```
1  K _ SQI = kurtosis (signal);
```

5.4.4 Entropy (E_{SQI})

Selvaraj et al.[75] found that entropy is a good indicator for PPG signal quality. Entropy quantifies how much the probability density function (PDF) of the signal differs from a uniform distribution, and thus provides a quantitative measure of the uncertainty present in the signal,[78] which is defined[79] as:

$$E_{SQI} = -\sum_{n=1}^{N} x[n]^2 \log_e \left(x[n]^2 \right) \tag{5.5}$$

where x signal is the raw PPG signal and N is the number of data points. This can be achieved in MATLAB as follows:

```
1  E_SQI = -sum((signal.^2) .* log(signal.^2));
```

5.4.5 Zero Crossing Rate (Z_{SQI})

This is the rate of sign changes in the processed signal (i.e. the rate at which the signal changes from positive to negative or back),[80] which is defined as:

$$Z_{SQI} = 1/N \sum_{n=1}^{N} \mathbb{I}\{y < 0\} \tag{5.6}$$

where y is the filtered PPG signal of length N, and $\mathbb{I}$, the indicator function $\mathbb{I}\{A\}$, is 1 if its argument A is true, and 0 otherwise. This step can be achieved using the following MATLAB code:

```
1  zeroCrossingNum = 0;
2  for i = 1:1:(length(signal)-1)
```

```
3              if   signal(i)*signal(i+1)<=0)
4                      zeroCrossingNum = zeroCrossingNum +1;
5              end
6 end
7 Z_SQI = zeroCrossingNum/length(signal);
```

5.4.6 Signal-to-Noise Ratio (N_{SQI})

This is a measure used in science and engineering that compares the level of a desired signal to the level of background noise. There are many ways to define the signal-to-noise ratio;[81] however, here the ratio of signal variance to the noise variance is used as follows:

$$N_{SQI} = \sigma_{signal}^2 / \sigma_{noise}^2 \qquad (5.7)$$

where σ_{signal} is the standard deviation of the absolute value of the filtered PPG signal (y) and σ_{noise} is the standard deviation of the y signal. This step can be achieved using the following MATLAB code:

```
1 std_signal = std(abs(filteredSignal));
       % filteredSignal is the PPG signal after filtered
2 std_noise = std(filteredSignal);
3 N_SQI = std_signal/std_noise;
```

5.4.7 Matching Systolic Detectors (M_{SQI})

Because different PPG algorithms are sensitive to different types of noise,[52] the comparison of how accurately multiple PPG systolic wave detectors isolate each event (such as a beat or noise artifact) provides one estimate of the level of noise in the PPG. A systolic wave detection algorithm can be used that is based on the first derivative with adaptive thresholds.[82] Alternatively, an algorithm can be used that is based on local maxima and minima.[83] These algorithms are referred to as Bing's algorithm and Billauer's algorithm. Bing's algorithm and Billauer's algorithm are easy to implement, and each algorithm approaches the PPG signal from a different perspective.[53] Matching of the algorithm outputs is defined as follows:

$$M_{SQI} = \left(S_{Bing} \cap S_{Billauer} \right) / S_{Bing} \qquad (5.8)$$

where S_{Bing} represents the systolic waves detected by Bing's algorithm, and $S_{Billauer}$ represents the systolic waves detected by Billauer's algorithm.

5.4.8 Relative Power (R_{sQI})

Frequency domain can be used to assess the PPG signal quality. Because most of the energy of the systolic and diastolic waves is concentrated within the 1–2.25 Hz[53] frequency band, the ratio of the power spectral density (PSD) in this band compared to the PSD in the overall signa ofl 0 – 8 Hz[53] provides a measure of the signal quality, which is defined as follows:

$$R_{SQI} = \sum_{f=1}^{2.25} PSD / \sum_{f=0}^{8} PSD \qquad (5.9)$$

Where the PSD was calculated using Welch's method, and the following MATLAB code can be used to calculate the relative power SQI:

```
1  signal=signal-mean(signal); %remove mean
2  NFFT = max(256,2^nextpow2(length(Wave)));
3  Fs = 125;    % Sampling frequency
4  %        Welch Method
5  [pxx,f] = pwelch(signal,length(signal),length
       (signal)/2,(NFFT*2)-1,Fs);
6
7  F1 = [1 2.25];
8  F2 = [0 8];
9
10
11 powerF1 = trapz(f(f>=F1(1)&f<=F1(2)),pxx(f>=F1(1)
       &f<=F1(2)));
12 powerF2 = trapz(f(f>=F2(1)&f<=F2(2)),pxx(f>=F2(1)
       &f<=F2(2)));
13
14 R_SQI = powerF1/powerF2;
```

5.5 SUMMARY

Current evaluations of SQIs for PPG signals are limited and lack thorough annotation. Consequently, comparing existing SQIs based on the current standard annotation provides an incomplete assessment of PPG signals. A more complete methodology for the annotation of PPG signals is provided. This chapter highlights the complexity of and the relationship between the annotation process and SQI assessment. Based on the analysis provided in Elgendi (2016),[84] skewness, which is the optimal SQI, can potentially be used to improve the diagnosis and monitoring of abnormalities, such as hypertension.

Mobile devices used at the point of care and that are often subject to noise would benefit from utilizing the SQI in the applicable software/app, as it will facilitate the recording of only high-quality signals. Implementing the optimal SQI on PPG-based mobile technologies is the first step towards reliable screening and monitoring solutions in settings where medical expertise is scarce, such as remote rural areas and developing countries. Pulse oximetry is increasingly becoming a go-to solution; it has multiple uses in health care settings and other off-site areas where there are patients. By helping to build a smart software/app that enables users to collect only high-quality signals, we are one step closer to increasing the accuracy of diagnoses and improving the quality of care.

CHAPTER 6

PPG Feature Extraction

PPG signals hold valuable information that can provide insight into not only the screening of diseases, but also diagnosis. This chapter gives an overview of the PPG features and how to extract them.

6.1 LEARNING OBJECTIVES

The learning objectives of this chapter are to:

- Learn about feature localization and the use of each feature in clinical applications

- Develop an understanding of how to use MATLAB code to localize PPG features

6.2 OVERVIEW OF PPG FEATURES

Figure 6.1 shows commonly extracted PPG features from different PPG measurement sites presented in studies conducted between 2010 and 2019. Many PPG features can be extracted, and this chapter attempts to clarify these features. It is clear from Figure 6.1 that the features extracted from PPG measured from the fingertip are the features most widely used and validated compared to other anatomical sites.

6.3 FEATURES OF PPG WAVEFORMS

The appearance of the PPG pulse is commonly divided into two phases: the anacrotic phase, which is the rising edge of the pulse, and the catacrotic phase, which is the falling edge of the pulse. The first phase is primarily concerned with systole, and the second phase with diastole and wave

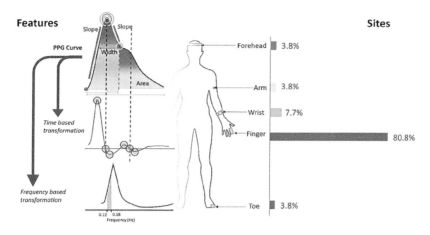

FIGURE 6.1 Overview of PPG features extracted from different PPG measurement sites in studies conducted between January 2010 and January 2019.[85]

reflections from the periphery. A dicrotic notch, shown in Figure 6.2, is usually seen in the catacrotic phase of subjects with healthy compliant arteries. A number of fiducial points extracted from the PPG waveform have been described in the literature; however, there is inconsistency in naming or labeling these points. Figure 6.2 shows a systematic, clear naming and labeling of the main fiducial points in the PPG waveform and its derivatives. Note that fiducial points are called "features", combinations of them, or even all of them, could be used to generate more sophisticated features such as time intervals, power, entropy, and soon.

6.3.1 Systolic Amplitude

Figure 6.2, displays the systolic amplitude (x), which is an indicator of the pulsatile changes in blood volume caused by arterial blood flow around the measurement site.[87,88] Systolic amplitude has been linked to stroke volume.[89] Dorlas and Nijboer found that systolic amplitude is directly proportional to local vascular distensibility over a remarkably wide range of cardiac outputs.[90] It has also been suggested that systolic amplitude is a potentially more suitable measure than the pulse arrival time for estimating continuous blood pressure.[91]

6.3.2 Pulse Width

The pulse width in the PPG wave is shown in Figure 6.3. Awad et al.[92] used the pulse width as the pulse width at the half height of the systolic peak. They have suggested that the pulse width correlates better with the systemic vascular resistance than with the systolic amplitude.

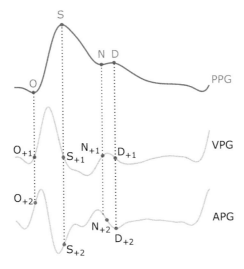

FIGURE 6.2 Demonstration of the photoplethysmogram (PPG) signal and its four waveforms [O(nset), S(ystolic), N(otch), and D(iastolic)].[86] *Notes:* The four PPG waveforms are mapped on the velocity photoplethysmogram (VPG) and the acceleration photoplethysmogram (APG) signals. The subscript +1 indicates the location of the PPG waveform on the first derivative of the PPG signal (i.e. VPG), while the +2 subscript indicates the location of the PPG waveform on the second derivative of the VPG signal (i.e. APG).

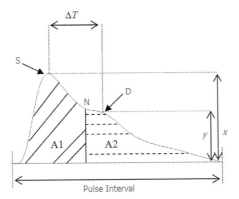

FIGURE 6.3 A typical waveform of the PPG and its characteristic parameters.

6.3.3 Pulse Area

The pulse area is measured as the total area under the PPG curve. Seitsonen et al.[93] found that the pulse area's response to a skin incision differs between movers and non-movers. Wang et al.[94] divided the pulse area into two sub-areas at the dicrotic notch; the ratio of these two areas, as shown in Figure

6.3, can be used as an indicator of total peripheral resistance. This ratio is called the "inflection point area ratio" (*IPA*) and is defined as follows:

$$IPA = A2/A1 \qquad (6.1)$$

6.3.4 Peak-to-Peak Interval

The distance between two consecutive systolic peaks is referred to as the peak–peak interval, as shown in Figure 6.3. The R–R interval in the ECG signal correlates closely with the Peak–Peak interval APG signal, as both represent a completed heart cycle. The peak–peak interval in PPG signals has been used to detect the heart rate and heart rate variability.[95–97] The peak–peak interval in PPG signals has been used to detect the heart rate and heart rate variability

6.3.5 Pulse Interval

The pulse interval is the distance between the beginning and the end of the PPG waveform, as shown in Figure 6.3. The pulse interval is typically used when the diastolic peaks are more clear and easier to detect compared to the systolic peak. Poon et al.[98] suggested that ratio of the pulse interval in relation to its systolic amplitude could provide an understanding of the properties of a person's cardiovascular system. In 2008, Lu et al.[62] compared heart rate variability (HRV) using the PPG pulse interval with the HRV using R–R intervals in ECG signals. Their results demonstrated that the HRV in PPG and ECG signals is highly correlated. They strongly suggested that PPG signals could be used as an alternative measurement of HRV.

6.3.6 Augmentation Index

The augmentation pressure (AG) is the measure of the contribution that the wave reflection makes to the systolic arterial pressure. It is obtained by measuring the reflected wave from the periphery to the center. Reduced compliance of the elastic arteries causes an earlier return of the "reflected wave", which arrives in systole rather than in diastole, causing a disproportionate rise in systolic pressure and an increase in pulse pressure, with a consequent increase in left ventricular afterload, a decrease in diastolic blood pressure, and impaired coronary perfusion. Takazawa et al.[12] defined the augmentation index (*AI*) as the ratio of y to x as follows:

$$AI = y/x \qquad (6.2)$$

As shown in Figure 6.3, y is the height of the late systolic peak and x is the early systolic peak in the pulse. Padilla et al.[4] used the RI as a reflection index as follows:

$$RI = y / x \qquad (6.3)$$

Rubins et al.[99] used the reflection index and introduced an alternative augmentation index as follows:

$$AI = (x - y)/x \qquad (6.4)$$

6.3.7 Large Artery Stiffness Index

The systolic component of the waveform arises mainly from a forward-moving pressure wave transmitted along a direct path from the left ventricle to the finger. The diastolic component arises mainly from pressure waves transmitted along the aorta to the small arteries in the lower body, from where they are then reflected back along the aorta as a reflected wave which then travels to the finger. The upper limb provides a common conduit for both the directly transmitted pressure wave and the reflected wave and, therefore, has little influence on their relative timing. As shown in Figure 6.3, the time delay between the systolic and diastolic peaks (or, in the absence of a second peak, the point of inflection) is related to the transit time of pressure waves from the root of the subclavian artery to the apparent site of reflection and back to the subclavian artery. This path length can be assumed to be proportional to the subject height (h). Therefore, Millasseau et al.[3] formulated an index of the contour of the PPG (SI) that relates to large artery stiffness.

$$SI = h / \Delta T \qquad (6.5)$$

Timing of the discrete PPG components has been examined to formulate an index of the contour, as it relates to large artery stiffness SI. The time delay between the systolic and diastolic peaks decreases with age as a consequence of increased large artery stiffness and the increased pulse wave velocity of pressure waves in the aorta and large arteries.[100] Millasseau et al. proved that the SI increases with age.[3] In order to facilitate the interpretation of the original PPG waves, Ozawa differentiated the PPG signals to enable analysis of the PPG wave contour.[12]

6.4 FEATURES OF VPG SIGNALS

The velocity photoplthysmogram (VPG) waveform has been found to have more valuable information in recent research, and has several components that have not yet been officially named. Based on our knowledge, there is no commonly used term to refer to the VPG components, perhaps limiting investigation on this topic. It is therefore necessary to refer to specific terminology when discussing VPG components: we suggest the use of *w*, *x*, *y*, and *z* (as shown in Figure 6.4) to refer, respectively, to the maximum slope peak in the systolic of the VPG waveform, the local minima slope in the systolic of the VPG waveform, the global minima slope in the systolic of the VPG waveform, and the maximum slope peak in the diastolic of the VPG waveform. Note that we use small italic letters for the VPG waveform components because each is the result of applying a derivative.

6.4.1 Diastolic Point

Millasseau et al.[3] defined the diastolic point feature as the point at which the first derivative of the waveform is closest to zero (Z point), shown in Figure 6.4. This feature can be used for heart rate measurement.

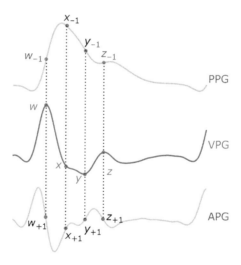

FIGURE 6.4 Demonstration of the velocity photoplethysmogram (VPG) signal and its four waveforms (*w*, *x*, *y*, and *z*).[86] *Notes*: The four VPG waveforms are mapped on the photoplethysmogram (PPG) and the acceleration photoplethysmogram (APG) signals. The subscript 1 indicates the location of the VPG waveform on the first integral of the VPG signal (i.e. PPG), while the subscript +1 indicates the location of the VPG waveform on the first derivative of VPG signal (i.e. APG). Note that the negative sign refers to the mathematical integration, while the positive sign refers to the mathematical derivation.

6.4.2 ΔT Calculation

The ΔT calculation is the peak-to-peak time feature related to the time taken for the pressure wave to propagate from the heart to the periphery and back. The time between the systolic and diastolic peaks is ΔT. The definition of ΔT depends on the PPG waveform, as its contour varies with different subjects. When there is a second peak, ΔT is defined as the time between the two maxima. However, in some PPG waveforms there is no clear second peak. In this case, ΔT is defined as the time between the peak of the waveform and the inflection point on the down slope of the waveform, which is a local maximum of the first derivative. This feature can be used to assess artery stiffness.

6.4.3 Crest Time Calculation

The crest time (CT) calculation feature is the time from the foot of the PPG waveform to its peak. Alty et al.[5] developed a method to categorize subjects as having either high or low pulse wave velocity (equivalent to a high or low risk of cardiovascular disease) using features extracted from the PPG. They found that the peak-to-peak time (ΔT), crest time (CT), and stiffness index ($SI = h/\Delta T$) were the best features for accurate classification of cardiovascular disease using the first derivative of the PPG. They used sets of these features for classification. Using a support vector machine-based classifier, they achieved a classification result of 87.5%. This feature can be used for cardiovascular disease classification.

6.5 FEATURES OF APG SIGNALS

The second derivative of the PPG signal is more commonly used than the first derivative for PPG signal analysis, as mainly contains useful features, and it reflects the acceleration of the blood flow. Every heartbeat in the APG signal consists of five waveforms: a, b, c, d, and e waveforms, as shown in Figure 6.5. The first waveforms (a, b, c, and d) represent a complete cycle of systole, while the e wave represents the onset of the diastole. Given there are already commonly used terms to refer to these waveforms, there is no need to rename or suggest other names/terminologies. Note that we use small italic letters for the APG waveforms because each is the result of applying a derivative.

6.5.1 a, b, c, d, and e Waves

As shown in Figure 6.5, the waveform of the APG includes four systolic waves and one diastolic wave: the a wave (early systolic positive wave), b wave (early systolic negative wave), c wave (late systolic reincreasing

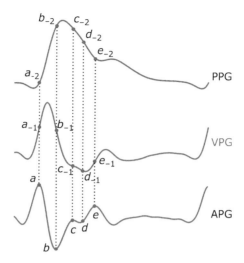

FIGURE 6.5 Demonstration of the acceleration photoplethysmogram (APG) signal and its five waveforms (*a*, *b*, *c*, *d*, and *e*).[86] *Notes*: The five APG waveforms are mapped on the photoplethysmogram (PPG) and the velocity photoplethysmogram (VPG) signals. The subscript 1 indicates the location of the VPG waveform on the first integral of the APG signal (i.e. VPG), while the subscript 2 indicates the location of the APG waveform on the second integral of the APG signal (i.e. PPG). Note that the negative sign refers to the mathematical integration.

wave), *d* wave (late systolic decreasing wave) and *e* wave (early diastolic positive wave). The *e* wave represents the dicrotic notch, as shown in Figure 6.5. Its location corresponds to the closure of the aortic valve and subsequent retrograde blood flow, and the site can be used to monitor cardiac function.[101] The height of each wave is measured from the baseline, with the values above the baseline being positive and those beneath it negative. The ratios of the height of each wave to that of the *a* wave (*b/a*, *c/a*, *d/a*, and *e/a*) are usually used for wave analyses.[67] The second derivative of the finger PPG waveform is used to stabilize the baseline, and to enable the individual features to be visualized and detected easily.

6.5.2 Ratio *b/a* Index

Takazawa et al.[12] demonstrated that the *b/a* ratio index reflects increased arterial stiffness; hence, the *b/a* ratio increases with age. Imanaga et al.[13] provided direct evidence demonstrating that the magnitude of the *b/a* of the APG is related to the distensibility of the peripheral artery. It is also suggested that the magnitude of *b/a* is a useful non-invasive index of atherosclerosis and altered arterial distensibility. Aiba et al.[102] discussed the parameter $-b/a$ in the exposure group dose dependently decreased with increases in length of working career (duration of exposure to lead and blood lead concentration (Pb-B). The parameter $-b/a$ significantly decreased in subjects with working careers of five years or more and in subjects whose Pb-B was $40\mu g/100ml$ or more. Otsuka et al.[103] found that the b/a index is positively correlated to the Framingham risk score. The Framingham risk score has been used to estimate individual risk of cardiovascular heart disease. Their results suggest that the b/a index could contribute to the discrimination of high-risk subjects for cardiovascular heart disease. Baek et al.[14] confirmed that the b/a ratio increases with age. Simek et al.[20] found that the b/a index discriminates independently between subjects with essential hypertension and healthy controls. Also, Zhang et al.[104] found the b/a ratio is associated with hypertension.

6.5.3 Ratio c/a Index

Takazawa et al.[12] demonstrated that the ratio c/a index reflects decreased arterial stiffness; hence, the index decreases with age. This index was also used by Simek et al. (2005)[20], who found that the c/a index distinguishes subjects with essential hypertension from healthy control subjects. Baek et al.[14] found that the c/a ratio decreases with age in the same way as the b/a ratio described in Section 6.5.3.

6.5.4 Ratio d/a Index

In 1998, Takazawa et al.[12] demonstrated that the d/a ratio index reflects decreased arterial stiffness; hence, the d/a ratio decreases with age. Moreover, they found the $-d/a$ ratio is a useful index not only for the evaluation of vasoactive agents, but also for left ventricular afterload. Baek et al.[14] confirmed that the d/a ratios decreases with age.

6.5.5 Ratio e/a Index

Takazawa et al.[12] demonstrated that an increase of the e/a ratio index reflects decreased arterial stiffness;hence, the e/a ratio decreases with age. Baek et al.[14] confirmed that the e/a ratio decreases with age.

6.5.6 Ratio $(b - c - d - e)/a$ Index

Takazawa et al.[12] found that the $(b - c - d - e)/a$ index increases with age and may be useful for the evaluation of vascular aging and for the screening of arteriosclerotic disease. Kimura et al.[15] calculated the vascular age as $45.5 \times (b - c - d - e)/a + 65.9$ years old.

6.5.7 Ratio $(b - e)/a$ Index

Baek et al.[14] suggested using the $(b - e)/a$ ratio index as the APG aging index when the c and d waves are missing, instead of using $(b - c - d - e)/a$.

6.5.8 Ratio $(b - c - d)/a$ Index

Ushiroyama et al.[19] reported that patients with a sensation of coldness showed an improvement in their $(b - c - d)/a$ index on treatment with a herbal supplement.

6.5.9 Ratio $(c + d - b)/a$ Index

Sano et al.[105] proposed a more comprehensive aging index $(c + d - b)/a$ and showed that it increases with age. Sano et al. distinguished seven main categories of APG signals depending on the waveform, as shown in Figure 6.6.

6.5.10 *aa* Interval

The R–R interval in the ECG signal correlates closely with the *aa* interval in the APG signal, as both represent a completed heart cycle. In 2007, Taniguchi et al.[21] used the *aa* interval instead of the *RR* interval to evaluate stress in surgeons. In 2010, Elgendi et al. calculated the heart rate and heart rate variability from the APG signals.[49,50]

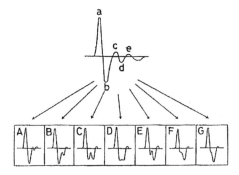

FIGURE 6.6   APG waveforms and types of photoplethysmogram.[106] *Notes:* There are different types of APG waveform. The first APG waveform A (far left) refers to good circulation, whereas the amplitude of the *b* wave is lower than that of the *c* wave. The last APG waveform G (far right) refers to distinctively bad circulation, whereas the amplitude of the *c* wave is lower than that of the *b* wave.

6.5.11 APG Beat Waveform

The shape of the APG waveform has been sorted into seven categories, referred to as A to G, shown in Figure 6.6. Type A is often observed in healthy young people and indicates good circulation. While, Types D–G are often observed in patients with cerebrovascular disease, ischemic heart disease, breast tumor, and uterine diseases. The changes from D to G reflect the developmental stages of the disease. Nousou et al.[20] developed a diagnostic system using APG and self-organizing maps (SOM). The original APG signal was adjusted in order to be classified correctly by the SOM. The *b* and the *d* wave were shifted along the time axis. A similar classification was used in Sano et al.,[105] as shown in Figure 6.6.

6.5.12 Segment of APG Signal

Tokutaka et al.[25] also developed a diagnostic tool to describe the general state of a person's health. They used the first section of the APG signal after the *a* peak in combination with SOM. Their approach was similar to that of Nousou et al.[20]

6.5.13 Chaos Attractor

Iokibe et al.[107] used the APG of healthy subjects and of patients with diseases varying from a common cold to pneumonia, intracerebral hemorrhage, and acute poisoning. Their aim was to find an indicator for the seriousness of the disease and the developmental stage of the disease and they applied chaos theory to the APG signals to develop their approach. Fujimoto et al.[22] proposed a criterion that combines two evaluations based on chaos theory: the trajectory parallel measurement method, and the size of the neighboring space in the chaos attractor to diagnose stress using the APG.

6.5.14 MATLAB Functions for Features Extraction

The detection of all waveforms mentioned depends on the PPG application. However, we can try simple functions that can be used as initial steps during the development of the desired algorithm.

An interesting MATLAB function *isoutlier*, which can be used for detecting events in PPG, VPG, and APG signals, and is defined as follows:

```
1 TF = isoutlier(A,method) %specifies a method for
      detecting outliers. For example , isoutlier
      (A, 'mean ') returns true for all elements more
      than three standard deviations from the mean
```

By default in MATLAB, an outlier is a value that is more than three scaled median absolute deviations (MAD) away from the median. This can be:

```
1 >> % Simulating a segment of APG signal and the
     traget is detecting the
2 >> A = [0.57 0.59 0.60 0.1 0.59 0.58 0.57 0.58 0.3
     0.61 0.62 0.60 0.62 0.58 0.57];
3 TF = isoutlier(A,'mean')
4
5 TF =
6
7   1 15 logical array
8
9   0   0   0   0   0   0   0   0   1   0   0   0   0
      0   0
```

Figure 6.7 shows the output of applying the *isoutlier* function using the method "mean". The results are useful, and can be used as an initial condition for optimizing the search for *a* waves. Note that no window was used to detect the *a* waves. Within the *isoutlier* function, a window length can be defined. For example, *isoutlier(A,'movmedian*', 100) returns true for all

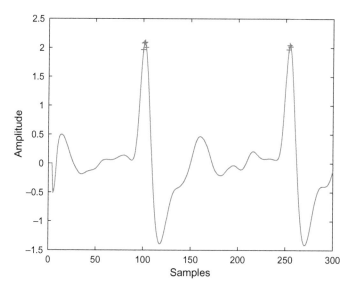

FIGURE 6.7 Localizing *a* waves in APG signals. *Notes*: Multiple outliers detected around each *a* wave can be used as an initial condition for optimizing the detection of *a* waves.

samples more than three local scaled MAD from the local median within a sliding window containing 100 samples.

Another function that can be used for detecting events in PPG signals is *islocalmax*, which is defined as:

```
1 TF = islocalmax(A) %returns a logical array whose
    elements are 1 ( true ) when a local maximum is
    detected in the corresponding element of an array ,
    table , or timetable.
```

After applying the *islocalmax* function to the same APG segment used in the previous example, the *e* wave was detected accurately; however, a few false events after the *e* wave were detected at sample 160, as shown in Figure 6.8.

After applying the *islocalmin* function to the same APG segment used in the previous examples, the *b* wave was detected accurately; however, a few false events before and after the *b* wave were detected at sample 120, as shown in Figure 6.9.

A generic algorithm for detecting events in PPG and APG signals are discussed in Chapter 7.

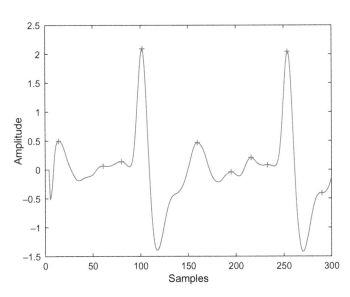

FIGURE 6.8 Localizing *c* and *e* waves in APG signals. *Notes:* In this example, the *islocalmax* was able to detect the *e* wave accurately.

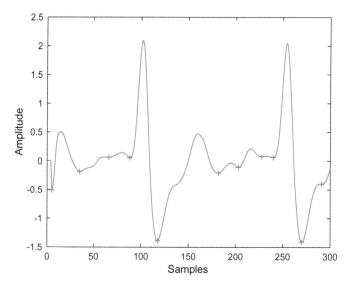

FIGURE 6.9 Localizing *b* and *d* waves in APG signals. *Notes*: In this example, the *islocalmin* was able to detect the *b* wave accurately.

6.5.15 MATLAB Code for Extracting 125 PPG Features

A comprehensive investigation of the characteristics of the PPG signal and its derivatives was conducted in Sections 6.4 and 6.5. For consistency and clarity, a summary and consistent names for PPG features are needed. The characteristics' features based on the morphological information are defined and include several types, such as time span (23 features), PPG amplitude (14 features), features of VPG and APG (10 features), waveform area (4 features), power area (15 features), ratio (43 features), and slope (16 features). A total of 125 features are defined and described, and these will be used in Section 12.7 in an example classifying normotensive and hypertensive states. Also, these features could be applied in other research. All of the wave names are presented in Figure 6.10.

6.5.15.1 Time Span

The time span features are expressed as their letters with a dash above. For example, the $\overline{SD}$ feature represents the time span from the systolic peak S to the diastolic peak D. A total of 23 time span features are defined.

6.5.15.2 Features of PPG Amplitude

The S, N, D, w_{-1}, a_{-2}, b_{-2}, c_{-2}, and so on features were defined in the PPG waveform. They represent the amplitude of the corresponding waveform from the PPG baseline. For example, the S amplitude feature represents the S height

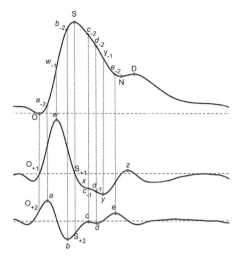

FIGURE 6.10 The definition of the characteristics in PPG and its derivatives. Note that all PPG, VPG and APG features shown in Figure 6.2, Figure 6.4 and Figure 6.5 are combined in this figure.

from the PPG baseline to the systolic peak S, and the N amplitude feature represents the N height from the PPG baseline to the diastolic notch N. Other features based on these features were also defined, such as N/S, w_{-1}/S, b_{-2}/S, c_{-2}/S, and so on. A total of 14 PPG amplitude features were defined.

6.5.15.3 Features of VPG and APG

The a, b, c, d, and e features were defined in the APG waveform, and the w, x, y, and z features were defined in the VPG waveform. They represent the amplitude of the corresponding waveforms from the APG baseline and VPG baseline. Other features based on these features were also defined, such as b/a, c/a, d/a, e/a, $(b - c - d - e)/a$, and $(b - c - d)/a$. A total of 20 VPG and APG features were defined.

6.5.15.4 Waveform Area

The waveform AC component area features are expressed as their letters with a polyline on top. For example, the $\overset{\frown}{OS}$ feature represents the AC component area under the curve from the onset point O to the systolic peak S. A total of 4 waveform area features were defined.

6.5.15.5 Power Area

The waveform AC component power area features are expressed as their letters with a bracket above. For example, the $\overset{\frown}{OS}$ feature represents the quadratic sum of the curve point from the onset point O to the systolic peak S. A total of 15 waveform area features were defined.

6.5.15.6 Ratio

The ratio features are expressed directly as their ratio formulae. For example, the $\overline{OS}/\overline{OO}$ feature represents the ratio of the time span feature $\overline{OS}$ and the time span feature $\overline{OO}$, and the $\widehat{OS}/\widehat{OO}$ feature represents the ratio of the curve area feature $\widehat{OS}$ and the curve area feature $\widehat{OO}$. A total of 33 ratio features were defined.

6.5.15.7 Slope

The slope features are expressed as their letters with a tilde on top. For example, the $\widetilde{OS}$ feature represents the slope from the onset point O to the systolic peak S. A total of 16 slope features were defined.

6.5.15.8 Code for PPG Feature Calculation

The code for PPG feature calculation is as follows:

```
1    function [error _ code,ppg _ feature] = ppg _
         feature _ calculation(error_code,sample_time,
         PPG_Loc,VPG_Loc,APG_Loc,ppg,vpg,apg)
2    %%%%%%%%%%%%%%%%%%%%%%%%%%%%%%%%%%%%%%%%%%%%%%%%%%%
3    %             ppg_feature_calculation.m
4    % Defination and Calculation of PPG Features
5    %%%%%%%%%%%%%%%%%%%%%%%%%%%%%%%%%%%%%%%%%%%%%%%%%%%
6
7    %% Input:
8    %error_code————Extracted PPG feature points or
         not. ( 0 :YES, 1: NO)
9    %sample_time————The sample time of PPG signal
10   %PPG_Loc————The location of PPG feature points
11   %VPG_Loc————The location of VPG feature points
12   %APG_Loc————The location of APG feature points
13   %ppg————PPG waveform of two adjacent heartbeat
         cycles
14   %vpg————VPG waveform of two adjacent heartbeat
         cycles
15   %apg————APG waveform of two adjacent heartbeat
         cycles
16
17   %% Output:
18   %error_code————Updated error code for
         calculation. ( 0 :YES, 1: NO)
```

```
19      %ppg_feature———A Structure of PPG features
20
21      %% Dilive PPG feature point location to
            variables
22      num_O = PPG_Loc(1);
23      num_S = PPG_Loc(2);
24      num_N = PPG_Loc(3);
25      num_D = PPG_Loc(4);
26      num_O_next = PPG_Loc(5);
27
28      num_w = VPG_Loc(1);
29      num_y = VPG_Loc(2);
30      num_z = VPG_Loc(3);
31      num_w_next = VPG_Loc(4);
32
33      num_a = APG_Loc(1);
34      num_b = APG_Loc(2);
35      num_c = APG_Loc(3);
36      num_d = APG_Loc(4);
37      num_e = APG_Loc(5);
38      num_b2 = APG_Loc(6);
39
40      ppg_feature.Total = zeros(1,125);
41      if error_code == 0
42          %% PPG Feature Type 1: Time Span
43          Tm_Oa = (num_a - num_O)*sample_time;
44          Tm_Ow = (num_w - num_O)*sample_time;
45          Tm_Ob = (num_b - num_O)*sample_time;
46          Tm_OS = (num_S - num_O)*sample_time;
47          Tm_Oc = (num_c - num_O)*sample_time;
48          Tm_Oy = (num_y - num_O)*sample_time;
49          Tm_ON = (num_N - num_O)*sample_time;
50          Tm_OD = (num_D - num_O)*sample_time;
51          Tss = (num_w_next - num_w)*sample_time;
52          Tm_Sc = (num_c-num_S)*sample_time;
53          Tm_Sd = (num_d-num_S)*sample_time;
54          Tm_Se = (num_e-num_S)*sample_time;
55          Tm_SD = (num_D - num_S)*sample_time;
56          Tm_ND = (num_D - num_N)*sample_time;
57          Tm_bb2 = (num_b2 - num_b)*sample_time;
58          Tm_bc = (num_c-num_b)*sample_time;
59          Tm_bd = (num_d-num_b)*sample_time;
60          Tm_wb = (num_b-num_w)*sample_time;
61          Tm_wS = (num_S-num_w)*sample_time;
62          Tm_wc = (num_c-num_w)*sample_time;
63          Tm_wd = (num_d-num_w)*sample_time;
```

```
64          Tm_wz = (num_z - num_w)*sample_time;
65          Tm_ac = (num_c -num_a)*sample_time;
66
67        ppg_feature.TimeSpan(1) = Tm_Oa;
68        ppg_feature.TimeSpan(2) = Tm_Ow;
69        ppg_feature.TimeSpan(3) = Tm_Ob;
70        ppg_feature.TimeSpan(4) = Tm_OS;
71        ppg_feature.TimeSpan(5) = Tm_Oc;
72        ppg_feature.TimeSpan(6) = Tm_Oy;
73        ppg_feature.TimeSpan(7) = Tm_ON;
74        ppg_feature.TimeSpan(8) = Tm_OD;
75        ppg_feature.TimeSpan(9) = Tss;
76        ppg_feature.TimeSpan(10) = Tm_Sc;
77        ppg_feature.TimeSpan(11) = Tm_Sd;
78        ppg_feature.TimeSpan(12) = Tm_Se;
79        ppg_feature.TimeSpan(13) = Tm_SD;
80        ppg_feature.TimeSpan(14) = Tm_ND;
81        ppg_feature.TimeSpan(15) = Tm_bb2;
82        ppg_feature.TimeSpan(16) = Tm_bc;
83        ppg_feature.TimeSpan(17) = Tm_bd;
84        ppg_feature.TimeSpan(18) = Tm_wb;
85        ppg_feature.TimeSpan(19) = Tm_wS;
86        ppg_feature.TimeSpan(20) = Tm_wc;
87        ppg_feature.TimeSpan(21) = Tm_wd;
88        ppg_feature.TimeSpan(22) = Tm_wz;
89        ppg_feature.TimeSpan(23) = Tm_ac;
90
91        %% PPG Feature Type 2: Features of PPG
              Amplitude
92        AMS = ppg(num_S) - ppg(num_O);
93        Am_Oa = ppg(num_a) - ppg(num_O);
94        Am_Ow = ppg(num_w) - ppg(num_O);
95        Am_Ob = ppg(num_b) - ppg(num_O);
96        Am_Oc = ppg(num_c) - ppg(num_O);
97        Am_Oy = ppg(num_y) - ppg(num_O);
98        Am_OO2 = ppg(num_O_next) - ppg(num_O);
99        Am_OD = ppg(num_D) - ppg(num_O);
100       Am_ON = ppg(num_N) - ppg(num_O);
101       Am_NS = ppg(num_S) - ppg(num_N);
102       AI_ON_AMS = Am_ON/AMS;
103       AI_OD_AMS = Am_OD/AMS;
104       AI_NS_AMS = Am_NS/AMS;
105       AI_DS_AMS = (ppg(num_S)-ppg(num_D))/AMS;
106
107       ppg_feature.Amplitude(1) = AMS;
108       ppg_feature.Amplitude(2) = Am_Oa;
```

```
109        ppg_feature.Amplitude(3)  = Am_Ow;
110        ppg_feature.Amplitude(4)  = Am_Ob;
111        ppg_feature.Amplitude(5)  = Am_Oc;
112        ppg_feature.Amplitude(6)  = Am_Oy;
113        ppg_feature.Amplitude(7)  = Am_OO2;
114        ppg_feature.Amplitude(8)  = Am_OD;
115        ppg_feature.Amplitude(9)  = Am_ON;
116        ppg_feature.Amplitude(10) = Am_NS;
117        ppg_feature.Amplitude(11) = AI_ON_AMS;
118        ppg_feature.Amplitude(12) = AI_OD_AMS;
119        ppg_feature.Amplitude(13) = AI_NS_AMS;
120        ppg_feature.Amplitude(14) = AI_DS_AMS;
121
122        %% PPG Feature Type 3: Features of VPG and
              APG
123        w = vpg(num_w);
124        y = vpg(num_y);
125        z = vpg(num_z);
126        a = apg(num_a);
127        b = apg(num_b);
128        c = apg(num_c);
129        d = apg(num_d);
130        e = apg(num_e);
131        cc = vpg(num_c);
132        dd = vpg(num_d);
133        r_z_w = z/w;
134        r_y_w = y/w;
135        r_cc_w = cc/w;
136        r_dd_w = dd/w;
137        r_b_a = b/a;
138        r_c_a = c/a;
139        r_d_a = d/a;
140        r_e_a = e/a;
141        r_bcde_a =(b-c-d-e)/a;
142        r_bcd_a = (b-c-d)/a;
143
144        ppg_feature.VpgApg(1) = w;
145        ppg_feature.VpgApg(2) = y;
146        ppg_feature.VpgApg(3) = z;
147        ppg_feature.VpgApg(4) = a;
148        ppg_feature.VpgApg(5) = b;
149        ppg_feature.VpgApg(6) = c;
150        ppg_feature.VpgApg(7) = d;
151        ppg_feature.VpgApg(8) = e;
152        ppg_feature.VpgApg(9) = cc;
```

```
153        ppg_feature.VpgApg(10) = dd;
154        ppg_feature.VpgApg(11) = r_z_w;
155        ppg_feature.VpgApg(12) = r_y_w;
156        ppg_feature.VpgApg(13) = r_cc_w;
157        ppg_feature.VpgApg(14) = r_dd_w;
158        ppg_feature.VpgApg(15) = r_b_a;
159        ppg_feature.VpgApg(16) = r_c_a;
160        ppg_feature.VpgApg(17) = r_d_a;
161        ppg_feature.VpgApg(18) = r_e_a;
162        ppg_feature.VpgApg(19) = r_bcde_a;
163        ppg_feature.VpgApg(20) = r_bcd_a;
164
165        %% PPG Feature Type 4: Waveform Area
166        S_OO = area_calculate(ppg(num_O),num_O,num
               _O_next,ppg);
167        S_OS = area_calculate(ppg(num_O),num_O,num_
               S,ppg);
168        S_Oc = area_calculate(ppg(num_O),num_O,num_
               c,ppg);
169        S_ON = area_calculate(ppg(num_O),num_O,num_
               N,ppg);
170
171        ppg_feature.WavefromArea(1) = S_OO;
172        ppg_feature.WavefromArea(2) = S_OS;
173        ppg_feature.WavefromArea(3) = S_Oc;
174        ppg_feature.WavefromArea(4) = S_ON;
175
176        %% PPG Feature Type 5: Power Area
177        power_OS_ppg = power_area_
           calculate(ppg(num_O),num_O, num_S,ppg);
178        power_wS_ppg = power_area_calculate
               (ppg(num_O),num_w, num_S,ppg);
179        power_Sc_ppg = power_area_calculate
               (ppg(num_O),num_S, num_c,ppg);
180        power_Sd_ppg = power_area_calculate
               (ppg(num_O),num_S, num_d,ppg);
181        power_OS_vpg = power_area_calculate
               (0,num_O,num_S,vpg);
182        power_wS_vpg = power_area_calculate
               (0,num_w,num_S,vpg);
183        power_Sc_vpg = power_area_calculate
               (0,num_S,num_c,vpg);
184        power_Sd_vpg = power_area_calculate
               (0,num_S,num_d,vpg);
```

```
185     power_OS_apg = power_area_calculate
           (0,num_O,num_S,apg);
186     power_wS_apg = power_area_calculate
           (0,num_w,num_S,apg);
187     power_Sc_apg = power_area_calculate
           (0,num_S,num_c,apg);
188     power_Sd_apg = power_area_calculate
           (0,num_S,num_d,apg);
189     power_OO_ppg = power_area_calculate
           (ppg(num_O),num_O,num_O_next,ppg);
190     power_OO_vpg = power_area_calculate
           (0,num_O,num_O_next,vpg);
191     power_OO_apg = power_area_calculate
           (0,num_O,num_O_next,apg);
192
193     ppg_feature.PowerArea(1) = power_OS_ppg;
194     ppg_feature.PowerArea(2) = power_wS_ppg;
195     ppg_feature.PowerArea(3) = power_Sc_ppg;
196     ppg_feature.PowerArea(4) = power_Sd_ppg;
197     ppg_feature.PowerArea(5) = power_OS_vpg;
198     ppg_feature.PowerArea(6) = power_wS_vpg;
199     ppg_feature.PowerArea(7) = power_Sc_vpg;
200     ppg_feature.PowerArea(8) = power_Sd_vpg;
201     ppg_feature.PowerArea(9) = power_OS_apg;
202     ppg_feature.PowerArea(10) = power_wS_apg;
203     ppg_feature.PowerArea(11) = power_Sc_apg;
204     ppg_feature.PowerArea(12) = power_Sd_apg;
205     ppg_feature.PowerArea(13) = power_OO_ppg;
206     ppg_feature.PowerArea(14) = power_OO_vpg;
207     ppg_feature.PowerArea(15) = power_OO_apg;
208
209     %% PPG Feature Type 6: Ratio
210     r_Tm_Oa_Tss = Tm_Oa/Tss;
211     r_Tm_Ow_Tss = Tm_Ow/Tss;
212     r_Tm_Ob_Tss = Tm_Ob/Tss;
213     r_Tm_OS_Tss = Tm_OS/Tss;
214     r_Tm_Oc_Tss = Tm_Oc/Tss;
215     r_Tm_Oy_Tss = Tm_Oy/Tss;
216     r_Tm_ON_Tss = Tm_ON/Tss;
217     r_Tm_wz_Tss = Tm_wz/Tss;
218     r_Tm_SD_Tss = Tm_SD/Tss;
219     r_Tm_bb2_Tss = Tm_bb2/Tss;
220     r_Oa_AMS = Am_Oa/AMS;
221     r_Ow_AMS = Am_Ow/AMS;
```

```
222          r_Ob_AMS = Am_Ob/AMS;
223          r_Oc_AMS = Am_Oc/AMS;
224          r_Oy_AMS = Am_Oy/AMS;
225          r_OO2_AMS = Am_OO2/AMS;
226          S_NO2 = area_calculate(ppg(num_O),num_N,num
                _O_next,ppg);
227          IPA = S_NO2/S_ON;
228          PIR = ppg(num_S)/ppg(num_O);
229          r_SOS_SOO = S_OS/S_OO;
230          r_SOc_SOO = S_Oc/S_OO;
231          r_SON_SOO = S_ON/S_OO;
232          r_OS_OO_ppg = power_OS_ppg/power_OO_ppg;
233          r_wS_OO_ppg = power_wS_ppg/power_OO_ppg;
234          r_Sc_OO_ppg = power_Sc_ppg/power_OO_ppg;
235          r_Sd_OO_ppg = power_Sd_ppg/power_OO_ppg;
236          r_OS_OO_vpg = power_OS_vpg/power_OO_vpg;
237          r_wS_OO_vpg = power_wS_vpg/power_OO_vpg;
238          r_Sc_OO_vpg = power_Sc_vpg/power_OO_vpg;
239          r_Sd_OO_vpg = power_Sd_vpg/power_OO_vpg;
240          r_OS_OO_apg = power_OS_apg/power_OO_apg;
241          r_wS_OO_apg = power_wS_apg/power_OO_apg;
242          r_Sc_OO_apg = power_Sc_apg/power_OO_apg;
243          r_Sd_OO_apg = power_Sd_apg/power_OO_apg;
244
245          ppg_feature.Ratio(1) = r_Tm_Oa_Tss;
246          ppg_feature.Ratio(2) = r_Tm_Ow_Tss;
247          ppg_feature.Ratio(3) = r_Tm_Ob_Tss;
248          ppg_feature.Ratio(4) = r_Tm_OS_Tss;
249          ppg_feature.Ratio(5) = r_Tm_Oc_Tss;
250          ppg_feature.Ratio(6) = r_Tm_Oy_Tss;
251          ppg_feature.Ratio(7) = r_Tm_ON_Tss;
252          ppg_feature.Ratio(8) = r_Tm_wz_Tss;
253          ppg_feature.Ratio(9) = r_Tm_SD_Tss;
254          ppg_feature.Ratio(10) = r_Tm_bb2_Tss;
255          ppg_feature.Ratio(11) = r_Oa_AMS;
256          ppg_feature.Ratio(12) = r_Ow_AMS;
257          ppg_feature.Ratio(13) = r_Ob_AMS;
258          ppg_feature.Ratio(14) = r_Oc_AMS;
259          ppg_feature.Ratio(15) = r_Oy_AMS;
260          ppg_feature.Ratio(16) = r_OO2_AMS;
261          ppg_feature.Ratio(17) = IPA;
262          ppg_feature.Ratio(18) = PIR;
263          ppg_feature.Ratio(19) = r_SOS_SOO;
264          ppg_feature.Ratio(20) = r_SOc_SOO;
265          ppg_feature.Ratio(21) = r_SON_SOO;
266          ppg_feature.Ratio(22) = r_OS_OO_ppg;
```

```
267    ppg_feature.Ratio(23) = r_wS_OO_ppg;
268    ppg_feature.Ratio(24) = r_Sc_OO_ppg;
269    ppg_feature.Ratio(25) = r_Sd_OO_ppg;
270    ppg_feature.Ratio(26) = r_OS_OO_vpg;
271    ppg_feature.Ratio(27) = r_wS_OO_vpg;
272    ppg_feature.Ratio(28) = r_Sc_OO_vpg;
273    ppg_feature.Ratio(29) = r_Sd_OO_vpg;
274    ppg_feature.Ratio(30) = r_OS_OO_apg;
275    ppg_feature.Ratio(31) = r_wS_OO_apg;
276    ppg_feature.Ratio(32) = r_Sc_OO_apg;
277    ppg_feature.Ratio(33) = r_Sd_OO_apg;
278
279    %% PPG Feature Type 7: Slope
280    m_Sc_ppg = (ppg(num_c)-ppg(num_S))/((num_c -
           num_S)*sample_time);
281    m_Sd_ppg = (ppg(num_d)-ppg(num_S))/((num_d -
           num_S)*sample_time);
282    m_bS_ppg = (ppg(num_S)-ppg(num_b))/((num_S -
           num_b)*sample_time);
283    m_bc_ppg = (ppg(num_c)-ppg(num_b))/((num_c -
           num_b)*sample_time);
284    m_bd_ppg = (ppg(num_d)-ppg(num_b))/((num_d -
           num_b)*sample_time);
285    m_wS_ppg = (ppg(num_S)-ppg(num_w))/((num_S -
           num_w)*sample_time);
286    m_OS_ppg = (ppg(num_S)-ppg(num_O))/((num_S -
           num_O)*sample_time);
287    m_ab_ppg = (ppg(num_b)-ppg(num_a))/((num_b -
           num_a)*sample_time);
288    m_ab_apg = (apg(num_b)-apg(num_a))/((num_b -
           num_a)*sample_time);
289    m_bS_apg = (apg(num_S)-apg(num_b))/((num_S -
           num_b)*sample_time);
290    m_bc_apg = (apg(num_c)-apg(num_b))/((num_c -
           num_b)*sample_time);
291    m_bd_apg = (apg(num_d)-apg(num_b))/((num_d -
           num_b)*sample_time);
292    m_be_apg = (apg(num_e)-apg(num_b))/((num_e -
           num_b)*sample_time);
293    m_Sc_apg = (apg(num_c)-apg(num_S))/((num_c -
           num_S)*sample_time);
294    m_wS_apg = (apg(num_S)-apg(num_w))/((num_S -
           num_w)*sample_time);
295    m_OS_apg = (apg(num_S)-apg(num_O))/((num_S -
           num_O)*sample_time);
296
```

```
297          ppg_feature.Slope(1)  = m_Sc_ppg;
298          ppg_feature.Slope(2)  = m_Sd_ppg;
299          ppg_feature.Slope(3)  = m_bS_ppg;
300          ppg_feature.Slope(4)  = m_bc_ppg;
301          ppg_feature.Slope(5)  = m_bd_ppg;
302          ppg_feature.Slope(6)  = m_wS_ppg;
303          ppg_feature.Slope(7)  = m_OS_ppg;
304          ppg_feature.Slope(8)  = m_ab_ppg;
305          ppg_feature.Slope(9)  = m_ab_apg;
306          ppg_feature.Slope(10) = m_bS_apg;
307          ppg_feature.Slope(11) = m_bc_apg;
308          ppg_feature.Slope(12) = m_bd_apg;
309          ppg_feature.Slope(13) = m_be_apg;
310          ppg_feature.Slope(14) = m_Sc_apg;
311          ppg_feature.Slope(15) = m_wS_apg;
312          ppg_feature.Slope(16) = m_OS_apg;
313
314          %% Data Exclusion Stardard :
315          %If the amplitude of 'O' in next heartbeat
                cycle is more than 50% baseline , the
                error code will be set to 1
316          if 0.50 < abs(r_O2_AMS)
317             error_code = 1;
318             ppg_feature.Total = zeros(1,125);
319          else
320             ppg_feature.Total = [ppg_feature.
                   TimeSpan ppg_feature.Amplitude ppg_
                   feature.VpgApg ppg_feature.Wave
                   fromArea ppg_feature.PowerArea ppg_
                   feature.Ratio ppg_feature.Slope];
312          end
322       end
323    end
```

6.5.15.9 Heart Rate Variability

One of the most extracted features from PPG is heart rate variability (HRV), which is the variation in the time interval between heartbeats. It is measured by the variation in the beat-to-beat interval. The methods used to calculate HRV can be grouped as follows: time domain, frequency domain, nonlinear.

To calculate HRV, the beat-to-beat intervals must be extracted. For PPG, the method for the detection of beats is to extract the peak-to-peak intervals or the valley-to-valley intervals. Errors in the location of the peaks or valleys will result in errors in the calculation of the HRV. Therefore, it is important to extract the correct peak or valley before calculating the HRV parameters.

The example PPG signal we use in this chapter is the 'p027337-2195-12-04-20-59m' record from the matched subset of the MIMIC III waveform database, which was downloaded from the Physionet website. The segment we used to calculate HRV is from 2,700 seconds to 3,000 seconds in this record, a total of 5 minutes.

The following code shows how to get the beat to beat intervals.

```
1  s = PPG; % PPG is the 5 minutes segment of the
           record 'p027337-2195-12-04-20-59m'
2  Fs = 125; %sampling frequency of MIMIC Database
3  peakLocs = detect_peaks(PPG); % detect peaks using
           one of the proposed algorithms discussed above.
4  intervals = diff(peakLocs)/Fs;
```

6.5.15.10 Time Domain Methods

The HRV parameters in time domain are based on the beat-to-beat intervals, which are analyzed as follows:

- SDNN: the standard deviation of intervals. Often calculated over a 24-hour period. SDANN, the standard deviation of the average intervals calculated over short periods, usually 5 minutes

- RMSSD: the square root of the mean of the squares of the successive differences between adjacent intervals

- SDSD: the standard deviation of the successive differences between adjacent intervals

- NN50: the number of pairs of successive intervals that differ by more than 50 ms

- pNN50: the proportion of NN50 divided by the total number of intervals

- NN20: the number of pairs of successive intervals that differ by more than 20 ms

- pNN20: the proportion of NN20 divided by the total number of intervals

The code of calculated the HRV parameters above as following:

```
1 SDNN = std(intervals)*1000;    % the unit of SDNN is
     millisecond
2 diffNN = diff(intervals)*1000;  % the unit of diffNN
     is millisecond
3 RMSSD = rms(diffNN);
4 SDSD = std(diffNN);
5 NN50 = length(find(abs(diffNN)>50));
6 pNN50 = NN50/length(intervals);
7 NN20 = length(find(abs(diffNN)>20));
8 pNN20 = NN20/length(intervals);
```

The result of these HRV parameters are as follows:

```
1 SDNN = 7.857;
2 RMSSD = 10.094;
3 SDSD = 10.105;
4 NN50 = 0;
5 pNN50 = 0;
6 NN20 = 25;
7 pNN20 = 0.052;
```

6.5.15.11 Frequency Domain Methods

The frequency domain methods are based on the spectral power of beat-to-beat intervals in different frequency ranges. The widely used HRV parameters in frequency domain are:

- VLF: total spectral power of all intervals between 0 and 0.04 Hz

- LF: total spectral power of all intervals between 0.04 and 0.15 Hz

- HF: total spectral power of all intervals between 0.15 and 0.4 Hz

- LF/HF: the ratio of LF to HF

The following code shows how to calculate the frequency domain HRV parameters using the Welch method:

```
1  %% interplotion
2  f_interplotion = 4;
3  t = zeros(1,length(intervals));
4  for j = 1:1:length(intervals)
5          t(j) = sum(intervals(1:j));
6  end
7  t2 = t(1):1/f_interplotion:t(length(t)); %time
     values for interp.
8  y=interp1(t,intervals,t2','spline')';     %cubic
     spline interpolation
9
10 %% power spectral density
11 y1=y-mean(y); %remove mean
12 NFFT = max(256,2^nextpow2(length(y1))); % the
     number of FFT
13 [pxx,f] = pwelch(y1,length(t2),length(t2)/2,(N
     FFT*2)-1,f_interplotion); % using welch method
     to calculate the power spectrum
14
15 %% HRV
16 f_VLF = [0 0.04];      % frequency range of VLF
17 f_LF = [0.04 0.15];    % frequency range of LF
18 f_HF = [0.15 0.4];     % frequency range of HF
19 VLF= trapz(f(f>=f_VLF(1)&f<=f_VLF(2)),pxx(f>=f_
     VLF(1)&f<=f_VLF(2))) * 1e6; % the unit of VLF is
     ms^2
20 LF = trapz(f(f>=f_LF(1)&f<=f_LF(2)),pxx(f>=f_LF(1)
     &f<=f_LF(2))) *1e6;    % the unit of LF is ms^2
21 HF = trapz(f(f>=f_HF(1)&f<=f_HF(2)),pxx(f>=f_HF(1)
     &f<=f_HF(2))) *1e6;    % the unit of HF is ms^2
22 ratio_LF_HF = LF/HF;
23 %% plot
24 figure;
25 plot(f(f>=f_VLF(1)&f<=f_VLF(2)),pxx(f>=f_VLF(1)
     &f<=f_VLF(2)),'color','k');
26 hold on;
27 plot(f(f>=f_LF(1)&f<=f_LF(2)),pxx(f>=f_LF(1)
     &f<=f_LF(2)),'color','r');
28 plot(f(f>=f_HF(1)&f<=f_HF(2)),pxx(f>=f_HF(1)
     &f<=f_HF(2)),'color','b');
```

```
29 set(gca,'XTick',[0, 0.04, 0.15,0.4]);
30 set(gca,'XTicklabel',{'0','0.04','0.15','0.4'});
```

The function *trapz(x, y)* integrates *y* with respect to the coordinates or scalar spacing specified by *x*. Figure 6.11 shows the spectrum of the intervals. And the results of VLF, LF, and HF are:

```
1        VLF = 14.81;
2        LF = 9.22;
3        HF = 16.80;
4        ratio_LF_HF = 0.55;
```

6.5.16 Nonlinear Methods

Linear methods cannot be fully described the HRV. Therefore, nonlinear methods have been applied to HRV so as to capture the characteristics of the beat-to-beat variability in full. Commonly used HRV parameters in nonlinear methods are:

- SD1: the standard deviation of the Poincaré plot perpendicular to the line of identity

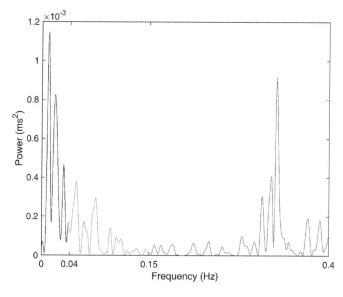

FIGURE 6.11 The power spectrum of the beat–beat intervals extracted from a PPG signal. *Notes:* The VLF is shown in black, the LF is shown in red, and the HF is shown in blue.

- SD2: the standard deviation of the Poincaré plot along to the line of identity

- SD2/SD1: the ratio of SD2 to SD1

- ApEn: approximate entropy of intervals

- SampEn: Sample entropy of intervals

6.5.16.1 Poincaré Plot

A Poincaré plot is a graph of NN(n) on the x axis versus NN(n + 1) (the succeeding NN interval) on the y axis, where NN stands for the beat-to-beat interval. This method takes a sequence of intervals and plots each interval against the subsequent interval. To characterize the shape of the plot, a common method has been to fit an ellipse to the plot. To fit the ellipse, a new set of axes were obtained by rotating the coordinate axis of the Poincaré plot by 45 degrees. SD1 is the standard deviation of the points in the new y axis. SD1 can be related to the time domain HRV as follows:

$$SD1 = SDSD / \sqrt{2} \tag{6.6}$$

SD2 is the standard deviation of the points in the new x axis. SD2 can be related to the time domain HRV as follows:

$$SD2 = \sqrt{2SDNN^2 - \frac{1}{2}SDSD^2} \tag{6.7}$$

The following code shows how to calculate the *SD1*, *SD2* and *SD2/SD1*:

```
1  diffNN = diff(intervals);
2  SD1 = sqrt(0.5*std(diffNN)^2)*1000;    % the unit of
     SD1 is ms
3  SD2 = sqrt(2*(std(intervals)^2) - (0.5*std
     (diffNN)^2))*1000; % the unit of SD2 is ms
4  ratio_SD2_SD1 = SD2/SD1;
5
6  %% draw Poincare plot
7  poincare_x = intervals(1:end-1)*1000;   %convert to ms
8  poincare_y = intervals(2:end)*1000;     %convert to ms
9  figure;
10 plot(poincare_x,poincare_y,'.');
11 hold on;
```

```
12 %draw ellipse
13 phi = pi/4; % The new coordinate system is
      established at 45 degree to the normal axis
14 new_x=poincare_x./cos(phi);         %translatex to
      new coordinate
15 center_new_x=mean(new_x);           %get the center
      of values along the new x axis
16 [cnx, cny]=deal(center_new_x*cos(phi),center_new_
      x*sin(phi)); % tranlsate center to new x , y
17 %    plot ([ min( poincare_x ) ,max( poincare_x ) ] ,
      [ min( poincare_y ),max( poincare_y ) ]) ;
18 ellipse_width=SD2;
19 ellipse_height=SD1;
20 theta = 0:0.01:2*pi;
21 x1 = ellipse_width*cos(theta);
22 y1 = ellipse_height*sin(theta);
23 X = cos(phi)*x1 - sin(phi)*y1;
24 Y = sin(phi)*x1 + cos(phi)*y1;
25 X = X + cnx;
26 Y = Y + cny;
27 plot(X,Y,'k-');
28 %plot SD1 and SD2 inside the ellipse
29 line_SD1=line([cnx cnx],[cny-ellipse_height
      cny+ellipse_height],'color','g');
30 rotate(line_SD1,[0,0,1],45,[cnx,cny,0]);
31
32 line_SD2=line([cnx-ellipse_width cnx+ellipse_
      width],[cny cny],'color','m');
33 rotate(line_SD2,[0,0,1],45,[cnx,cny,0]);
```

Figure 6.12 shows the Poincare plot. And the results of SD1 and SD2 as follows:

```
1 SD1 = 7.145;
2 SD2 = 8.510;
```

6.5.16.2 Approximate Entropy and Sample Entropy

Approximate entropy (ApEn) is a technique used to quantify the regularity and unpredictability of time series data fluctuations. The algorithm for approximate entropy is as follows:

- Form a time series of data $u(1)$, $u(2)$, ..., $u(N)$

- Fix intervariable m represents the length of the compared run of data, and a positive real number r specifies a filtering level

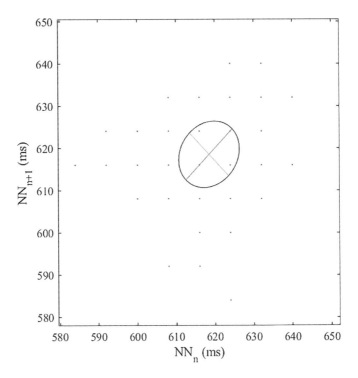

FIGURE 6.12 Poincaré plot of the beat–beat intervals extracted from a PPG signal. *Notes*: The green line represents SD1, while the magenta line represents SD2.

- Form a sequence of vectors $x(1)$, $x(2)$, ..., $x(N - m + 1)$ in R^m, real m-dimensional space defined by $x(i) = [u(1), u(2), ..., u(i + m - 1)]$

- Use the sequence $x(1)$, $x(2)$, ..., $x(N - m + 1)$ to construct, for each i, $1 \leq i \leq N - m + 1$

$$C_i^m = \left(\text{number of } x(j) \text{ such that } d\left[x(i), x(j) \right] \leq r \right) / \left(N - m + 1 \right) \qquad (6.8)$$

in which $d[x, x']$ is defined as:

$$d\left[x, x' \right] = \max_a \; u(a) - u'(a) \qquad (6.9)$$

The $u(a)$ are the m scalar components of x. d represents the distance between the vectors $x(i)$ and $x(j)$, given by the maximum difference in their respective scalar components. Note that j takes on all values, so the match provided when $i = j$ will be counted.

- The approximate entropy is defined as:

$$ApEn = \Phi^m(r) - \Phi^{m+1}(r) \qquad (6.10)$$

where the $\Phi^m(r)$ is defined as:

$$\Phi^m(r) = (N - m + 1)^{-1} \sum_{i=1}^{N-m+1} \log C^m(r) \qquad (6.11)$$

When calculating the *ApEn* of HRV, we used the option $m = 2$ and $r = 0.2std(intervals)$. The R2018a version of MATLAB includes an approximate entropy function in the Predictive Maintenance Toolbox. We can calculate ApEn as follows:

```
1 ApEn = approximateEntropy(intervals,'dim',2,'Radius',0.2
    *std(intervals));
```

The results of ApEn is: *ApEn* = 1.1102

Similar to ApEn, sample entropy(SampEn) was used to assess the complexity of time series signals. The algorithm of SampEn is as follows:

- Given a time series $u(1), u(2), \ldots, u(N)$

- Form a vector of length m, such that $X_m(i) = x_i, x_{i+1}, \ldots, x_{N+m-1}$

- select a distance function, such as the Chebyshev distance.

$$D(x,y) = \max_i (x_i - y_i) \qquad (6.12)$$

- The SampEn is defined as follows:

$$SampEn = -\log \frac{A}{B} \qquad (6.13)$$

where A = the number of vector pairs having $D[x_{m+1}(i), X_{m+1}(j)] < r$, and B = number of vector pairs having $D[x_m(i), X_m(j)] < r]$.

The parameter r is the tolerance. To calculate SampEn in HRV, we typically used $m = 2$ and $r = 0.2std(intervals)$.

There is a function to calculate the sample entropy in the MATLAB File Exchange (https://www.mathworks.com/matlabcentral/fileexchange/69381-sample-entropy). It can be downloaded from the File Exchange. The following code shows how to calculate the sample entropy:

```
1 SampEn = SampEn(2, 0.2*std(intervals), intervals);
```

The results of SampEn is: *SampEn* = 1.1374.

Note that SampEn produced a slightly different result compared with approximateEntropy.

6.5.17 Discussion

Photoplethysmography is advantageous as it is relatively low-cost, simple, and can be applied to portable technology that can be used in primary health care facilities and remote clinics. This review has described a number of features of PPG and its respective derivatives, and the potential application of each feature. Two PPG indices based on the original PPG signal were described: the augmentation index, and large artery stiffness index. Features extracted from the first derivative of the PPG were discussed: the diastolic point, the peak-to-peak time (ΔT), and the crest time. The first derivative of the PPG can also be used to calculate the augmentation index and the large artery stiffness index more accurately.

Most indices are based on the second derivative photoplethysmogram (APG), which provides more information than the first derivative of the PPG. The indices calculated from the APG waveform are reported to correlate closely with the distensibility of the carotid artery,[107] age,[12] blood pressure,[108] the estimated risk of coronary heart disease,[103] and the presence of atherosclerotic disorders.[18] Some of the APG indices have been calculated with different formulae. For example the aging index can be calculated as $(b - c - d - e)/a$, $(b - e)/a$ or $(c + d - b)/a$. A number of indices are reported to indicate vascular stiffness; the b/a index increases with increasing arterial stiffness, while the c/a, d/a and e/a indices decrease.

Determining which of the APG indices is the most informative is challenging; at times, the same feature is used as a measure of different, but potentially related, physiological variables. For example, the b/a ratio has been used as an indicator of arterial stiffness, hypertension, vascular aging, and risk of cardiovascular disease. In addition to cardiovascular risk factors, the APG has also been described as a potential diagnostic tool for other disorders, varying from a sensation of coldness[19] and stress experienced by surgeons[21] to exposure to lead.[103] pneumonia, intracerebral haemorrhage, and acute poisoning.[107] This has its origins in Eastern medicine, where the pulse is considered a very important diagnostic variable. Self-organizing maps and chaos theory have been applied to reveal the stage of a disease or the general state of health.[22,25,107] Currently, full understanding of the diagnostic value of the different features of a PPG signal is still to be achieved and more research is needed.

6.6 SUMMARY

Photoplethysmography technology has immense potential for diagnostic application. A common structure of any PPG diagnostic system consists of three stages: preprocessing, feature extraction, and diagnosis. Preprocessing and feature extraction were discussed to promote a thorough understanding of the PPG signal and its features.

The different artifact sources that can affect the PPG signal in the preprocessing stage have been described. These sources may be a power line interface, motion artifacts, low amplitude, or the existence of arrhythmia.

The characteristics of the PPG waveform and its derivatives have been clarified in relation to the feature extraction stage. Features of the PPG signal have been discussed. These features may be calculated based on the original signal, or on the first or second derivative of the PPG signal. Taking the first and second derivatives of PPG signals may help in detecting informative inflection points more accurately. Different features have been used as indicators for the same physiological variables. Several vascular stiffness and aging indices have been described, the most informative of which is currently not clear. Some features have been used as indicators of different, but potentially related, cardiovascular variables. Features of the second derivative of the PPG have also been described in the literature as indicators for the general state of health. There is no doubt that the HRV indices are the most common PPG features used in clinical assessment. However, the combination of all the features discussed has the potential to be applied to many other pathological studies.

CHAPTER 7

A Generic Method for Event Detection

This chapter discusses a generic method that can be used to detect events within PPG signals. The method is called "two event-related moving averages", and is referred to as TERMA.[109] The generalization and adaptability of TERMA make it attractive for detecting features in PPG signals. TERMA consists of five variables: once these variables are optimized in the training phase for detecting a certain event, they can be used as is in the testing phase to detect the required event.

7.1 LEARNING OBJECTIVES

The learning objectives of this chapter are to:

- Develop an understanding of a generic method for peaks and events within PPG signals

- Learn how to optimize parameters for detecting events

- Gain insight into how to develop a generic method that captures a specific event in a robust, efficient, and consistent manner

7.2 INTRODUCTION

Clinicians use biomedical signals, such as PPG signals, to screen and diagnose various cardiac abnormalities. Collecting these biomedical signals is relatively easy and inexpensive when compared to invasive alternatives[110] Therefore, the analysis of biomedical signals has been extensively investigated since the turn of the millennium. Many algorithms using a variety of mathematical formulae have been published to analyze biomedical signals; however, there is no generic methodology with a clear framework that can be used to analyze these signals. Such a generic methodology may provide physicians with greater insights about a patient's health through non-invasive measures.

A generic framework that has been well-established in the field of economics to analyze financial data is the use of two moving averages. A moving average is commonly used with time series data to smooth out short-term fluctuations and to highlight longer-term trends or cycles. The use of one moving average is a common analysis tool used by traders to identify trend directions.

Two moving averages have been used together to generate crossover signs.[111,112] A crossover occurs when a faster (shorter) moving average crosses a slower (longer) moving average,[111,112] and these crossovers are considered as buy and sell indicators. The use of two moving averages succeeds in detecting the critical events in trading. Looking at the NASDAQ composite index for the calendar year 2001, if the closing values are filtered, a large proportion of the day-to-day market variations can be removed. For example, with the use of two moving averages—the shorter with a 4-day window length and the longer with a 32-day window length— the number of remaining variations can be controlled. The moving average with the longer window length works as a threshold to the moving average with the shorter window length and, consequently, presents a crossover as an indicator of a critical event, as shown in Figure 7.1.

It is common practice in biomedical signal analysis to use the moving average as a filter. It is important to note that the moving average step has not been previously used in decision-making. For example, the moving average used in Matsuyama (2009),[113] Mattson et al. (2004),[114] and Oppenheim and Shafer (1989)[115] was only a filtering step; it was not used for decision-making (thresholding) in the same way as it is applied in economics. The implementation of the moving average can be highly numerically efficient (simple, fast, and with fewer calculations required). Therefore, the idea of using two moving averages is promising for analyzing biomedical signals.

Analyzing real-time biomedical signals collected by a battery-driven device needs to be fast and feasible in real time, despite the existing limitations in terms of memory and processor capability. The same holds for the ability to analyze large biomedical signals collected over one or more days.

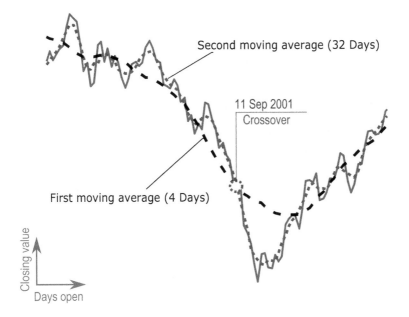

FIGURE 7.1 Filtered and unfiltered closing values of the NASDAQ composite index for the calendar year 2001. *Notes*: The dashed black line is the first moving average with a 4-day window length, and the dotted red line is the second moving average with a 32-day window length.

The main goal of this study is to produce a methodology that can be used for detecting different types of events in different types of biomedical signals using two event-related moving averages (TERMA). The window sizes of the moving averages depend on prior knowledge of the expected duration of the event to be detected. In this chapter, I will demonstrate and discuss how TERMA can be used to detect events in different research areas related to biomedical signals.

7.3 DATA USED

In this section, PPG and APG signals are used. Each biomedical signal has it is own unique set of features and events. A single PPG pulse signal consists of a systolic wave; a single APG heartbeat signal consists of *a, b, c, d,* and *e* waves. The databases used to detect these events are:

- For systolic wave detection in PPG signals: one annotated Heat-Stress PPG Database[53] consisting of 5,071 beats of 40 healthy, heat-acclimatized emergency responders (30 males and 10 females). The PPG data were collected at a sampling rate of 367 Hz, and the duration of each recording was 20 s. The data used in the training set were the PPG signals measured at rest, while the data used in the test set were the PPG signals measured after three simulated heat stress exercises.

- For *a*, *b*, *c*, *d* and *e* wave detection in APG signals: one annotated Heat-Stress PPG Database[113] consisting of 1,469 beats of 27 healthy volunteers (males). The PPG data were collected at a sampling rate of 200 Hz, and the duration of each recording was 20 s. The data used in the training set were the APG signals after 1 hour of exercise, while the data used in the test set consisted of the APG signals measured at rest and after 2 hours of exercise.

Note that Chapter 11 discusses some publicly available PPG databases that can be used for training and testing.

7.4 TERMA FRAMEWORK

In this section, a new, knowledge-based, numerically efficient, and robust method is proposed to detect main events in biomedical signals using the TERMA algorithm. The structure of the TERMA.[109] Algorithm is shown in Figure 7.2.

It is clear that prior knowledge of TERMA parameters supports decision-making in both stages, generating blocks of interest and thresholding. The more precise the prior knowledge, the higher the overall performance and accuracy of detection. The pseudocode of the TERMA detector is as follows:

```
function Detector(F_1, F_2, W_1, W_2, )
    x ← Bandpass(ECG, F_1-F_2)
    y ← x²
    MA_event←MA(y, W_1)
    MA_cycle←MA(y, W_2)
    z←ȳ
    α ← (β × z + MA_cycle)
    THR_1 ← MA_cycle + α
    for n = 1 to length(MA_event) do
        if MA_event[n] > THR_1 then
```

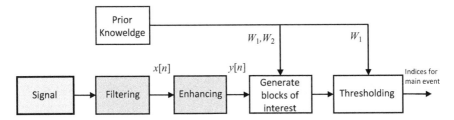

FIGURE 7.2 Flowchart of the two event-related moving averages (TERMA) algorithm for detecting the main event in a quasi-periodic signal.[109] *Notes:* The algorithm consists of six LEGO building bricks: signal, filtering, enhancing, generating blocks of interest, thresholding, and prior knowledge.

```
        BlocksOfInterest[n] ← 0.1
    else
        BlocksOfInterest[n] ← 0
    end if
end for
Blocks ← onset and offset from BlocksOfInterest
THR₂ ← W₁
for j = 1 to number of blocks do
    if width(Blocks[j]) ≥ THR₂ then
            Peaks ← max value with this block
    else
            Ignore this block
    end if
end for
end function
```

7.4.1 Prior Knowledge

When using TERMA, prior knowledge of the duration of the main events of the biomedical signals can assist in feature extraction and, thus, support the decision-making of the algorithm. Four parameters are required as prior knowledge: the frequency band (F_1–F_2), event-related durations W_1 and W_2, and the offset fraction (β). Usually, the TERMA prior knowledge for these parameters is not reported in the literature.

The literature has not yet reported this need for prior knowledge of the TERMA parameter values for all biological events . Therefore, it is recommended that a subset of the data is taken to determine the duration of the main events via an optimization step. In other words, the output of the optimization step will be used as prior knowledge for the rest of the dataset.

The optimization step has five decision variables: F_1, F_2, W_1, W_2 and β. The improvement in one objective will result in the worsening of at least one other objective, generating Pareto solutions.[114] Any change in these parameters affects the overall performance of the proposed algorithm. The five decision variables are interrelated and cannot be optimized in isolation. Our goal is to find the Pareto optimal point, within all possible Pareto solutions, for this multi-objective problem. An aggregate objective function, denoted by J, to combine two objective functions into a scalar function is defined as follows:

$$\max_{F_1,F_2,W_1,W_2,\beta} \quad J = \frac{1}{2}\left\{\left[\mathbf{SE}\left(F_1,F_2,W_1,W_2,\beta\right)\right]+\left[+\mathbf{P}\left(F_1,F_2,W_1,W_2,\beta\right)\right]\right\}$$

subject to

$$f_{1\min} \leq F_1 \leq f_{1\max} \;,$$
$$f_{2\min} \leq F_2 \leq f_{2\max} \;,$$
$$w_{1\min} \leq W_1 \leq w_{1\max} \;,$$
$$w_{2\min} \leq W_2 \leq w_{2\max} \;,$$
$$b_{\min} \leq \beta \leq b_{\max} \;,$$

where J is the overall accuracy, which is defined as the average of sensitivity (SE) and positive predictivity (+P). SE and +P are the two objective functions to be maximized jointly. The Pareto frontier is formed with solutions (the values of the five decision variables) that optimize them both. Finding the optimal Pareto point goes through a brute force search of all parameters; though time-consuming, once this has been achieved the optimal solution will be used as is for the implementation.

7.4.2 Bandpass Filter

Morphologies of normal and abnormal events in biosignals differ widely. Biosignals are often corrupted by noise from many sources; therefore, bandpass filtering is an essential first step for nearly all event detection algorithms. The purpose of bandpass filtering is to remove the baseline wander and high frequencies that do not contribute to detecting these events. A bandpass filter is typically used as a bidirectional Butterworth implementation.[115] It offers good transition band characteristics at low coefficient orders, which makes it efficient to implement.[115] All research was carried out using the TERMA method, a third-order Butterworth filter with a passband of F_1–F_2 Hz to remove baseline wander and high frequencies, where F_1 is the starting frequency and F_2 is the stopping frequency.

7.4.3 Signal Enhancement

The signal is squared point-by-point to enhance large values and boost high-frequency components using the following equation:

$$y[n] = (x[n])^2 \tag{7.1}$$

7.4.4 Generating Blocks of Interest

Blocks of interest are generated using two event-related moving averages. The first moving average, MA_{event}, is used to extract a specific event (within a cycle), while the second moving average, MA_{cycle}, extracts the cycle (regularly repeating events). Next, an event-related threshold is applied to the generated blocks to distinguish the blocks that contain the event peaks from the blocks that include noise. The purpose of the event moving average (MA_{event}) is to smooth out multiple peaks corresponding to the event length to emphasize and extract the event area:

$$MA_{event}[n] = \frac{1}{W_1}(y[n-(W_1-1)/2]+\ldots$$
$$+y[n]+\ldots+y[n+(W_1-1)/2]), \tag{7.2}$$

where W_1 is the approximate duration of a specific event, rounded to the nearest odd integer, and n is the number of data points. The value of W_1 is determined based on the prior knowledge discussed above.

The purpose of the cycle moving average (MA_{cycle}) is similar to that of the MA_{event}, but emphasizes the cycle area that contains the event of interest to be used as a threshold for the first moving average (MA_{event}):

$$MA_{cycle}[n] = \frac{1}{W_2}(y[n-(W_2-1)/2]+\ldots$$
$$+y[n]+\ldots+y[n+(W_2-1)/2]), \tag{7.3}$$

where W_2 is the approximate duration of a cycle (or heartbeat), rounded to the nearest odd integer, and n is the number of data points. The value of W_1 is determined based on the prior knowledge discussed above.

The blocks of interest are generated based on the two moving averages. In other words, applying the second moving average (MA_{cycle}) as a threshold to the first moving average (MA_{event}) produces blocks of interest, as shown in Figure 7.3. However, the use of MA_{cycle} without an added offset reduces the detection accuracy due to its sensitivity to a low signal-to-noise ratio (SNR).

Here, the SNR is defined as the ratio of the mean signal of a region of interest to its standard deviation,[116] which means that if the statistical mean of the signal increases, the SNR increases. This leads to introducing an offset based on the statistical mean of the signal as:

$$\alpha = \beta \times \bar{z} \qquad (7.4)$$

where β is the fraction of the $\bar{z}$ signal that needs to be removed, $\bar{z}$ is the statistical mean of the squared PPG signal z, and α is an offset for the threshold MA_{cycle} signal. Therefore, α refers to the offset, while β refers to the offset fraction.

In short, to increase the accuracy of detecting events in noisy biosignals, the dynamic threshold value THR_1 is calculated by offsetting the MA_{cycle} signal with α, as follows:

$$THR_1 = MA_{cycle}\left[n\right] + \alpha \qquad (7.5)$$

The blocks of interest are then generated by comparing the MA_{event} signal with THR_1. If a block is higher than THR_1, it is classified as a block of interest containing biosignal features (different events) and noise. By this stage, the blocks of interest are generated and stored in *Blocks[n]* as a square pulse (ones and zeros). Therefore, the next step is to reject the blocks that result from noise. The rejection should be related to the anticipated block width.

7.4.5 Thresholding

Here, blocks containing undesired events (or events that are of no interest) and noise are rejected using the new THR_2 threshold. By applying the THR_2 threshold, the accepted blocks contain only the required events:

$$THR_2 = W_1 \qquad (7.6)$$

As discussed, the threshold THR_2 equals W_1, which corresponds to the anticipated event width. If the block width equals the window size W_1, then the block contains an event. However, the event duration varies in terms of the durations of the abnormal events within the processed signal. Therefore, the condition is set to capture both normal and abnormal event

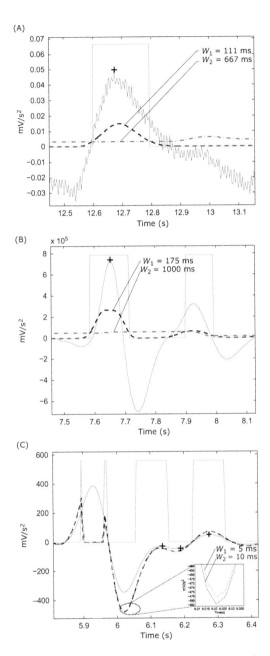

FIGURE 7.3 Demonstrating the effectiveness of TERMA in detecting events in biomedical signal.[109] *Notes*: (A) systolic wave detection; (B) *a* wave detection; (C) *c*, *d*, and *e* wave detection. The dashed black line is the MA_{event} with W_1, and the dotted red line is the MA_{cycle} with W_2. The peak of the investigated event is detected using TERMA (represented by a black plus sign) within the blocks of interest (represented by a green square pulse).

durations. Therefore, if a block width is greater than or equal to W_1, it is classified as an event. If not, the block is classified as an undesired event or noise.

7.4.6 Detecting Event Peak

The last stage is finding the maximum absolute value within each block, or the event peak.

7.5 RESULTS

The event detection algorithm is typically run using two statistical measures: SE and +P, where SE = TP/(TP + FN) and +P = TP/(TP + FP). Here, TP is the number of true positives (events detected as events), FN is the number of false negatives (events that have not been detected as events), and FP is the number of false positives (non-events detected as events). The SE reports the percentage of true events that were correctly detected by the algorithm. The +P reports the percentage of event detections that were true events.

7.5.1 Training Results

The training dataset for each detection problem is discussed in Section 7.3. A rigorous optimization using a brute force search of all parameters is conducted as follows:

- For systolic wave detection in PPG signals: the lower frequency resulted in a value from $f_{1min} = 0.5$ Hz to $f_{1max} = 1$ Hz, while the higher frequency resulted in a value from $f_{2min} = 7$ Hz to $f_{2max} = 15$ Hz. The window size of W_1 varied from $w_{1min} = 54$ ms to $w_{1max} = 111$ ms, whereas the window size of W_2 varied from $w_{2min} = 545$ ms to $w_{2max} = 694$ ms. The offset β was tested over the range $b_{min} = 0\%$ to $b_{max} = 10\%$.

- For a and b wave detection in APG signals: the lower frequency resulted in a value from $f_{1min} = 0.5$ Hz to $f_{1max} = 1$ Hz, while the higher frequency resulted in a value from $f_{2min} = 7$ Hz to $f_{2max} = 15$ Hz. The window size of W_1 varied from $w_{1min} = 100$ ms to $w_{1max} = 200$ ms, whereas the window size of W_2 varied from $w_{2min} = 1000$ ms to $w_{2max} = 1250$ ms. The offset β was tested over the range $b_{min} = 0\%$ to $b_{max} = 10\%$.

- For c, d and e wave detection in APG signals: the lower frequency varied from $f_{1min} = 0.5$ Hz to $f_{1max} = 1$, while the higher frequency varied from $f_{2min} = 4$ Hz to $f_{2max} = 10$ Hz. The window size of W_1 varied from $w_{1min} = 5$ ms to $w_{1max} = 25$ ms, whereas the window size of W_2 varied from $w_{2min} = 10$ ms to $w_{2max} = 15$ ms, while the range of β varied from $b_{min} = 0\%$ to $b_{max} = 10\%$.

The databases used in the optimization process contain abnormal rhythms, different event morphologies, heat stress signals, and low SNR signals. Several publications have listed the use of all files in the database, excluding the paced patients, segments, and certain beats.[120] However, in the optimization process, all records were used without excluding any segment or beat.

As we have multiple objectives, plotting the Pareto frontier (the objective space of possible Pareto solutions) cannot be achieved. Therefore, all Pareto solutions were sorted in descending order according to their overall accuracy (objective function J); thus, the first combination is considered the optimal Pareto solution.

After applying the multiple objective optimization step, the optimal Pareto solution for the detection of systolic waves in PPG signals was found to be $F_1 = 0.5$ Hz, $F_2 = 8$ Hz, $W_1 = 111$ ms, $W_2 = 667$ ms, and $\beta = 2\%$. The optimal solution for detecting a and b waves was $F_1 = 0.5$ Hz, $F_2 = 15$ Hz, $W_1 = 175$ ms, $W_2 = 1000$ ms, and $\beta = 0\%$; and the optimal solution for detecting c, d and e waves was found to be $F_1 = 0.5$ Hz, $F_2 = 7$ Hz, $W_1 = 5$ ms, $W_2 = 15$ ms, and $\beta = 0\%$.

7.5.2 Testing

An optimal event detector is obtained from the training phase. We can then test each detector on its testing dataset "straight out of the box" without any tuning. In other words, the algorithm's parameters (F_1, F_2, W_1, W_2 and β) do not need to be trained in a real-world application for every subject. The parameters are optimized on a large training set; thus, the robustness of the algorithm can be examined against different databases with different sampling frequencies, and the biosignals can be collected by different doctors in dissimilar conditions.

The performance of the TERMA-based detection algorithm on the testing datasets can be summarized as follows:

- For systolic wave detection in PPG signals: the TERMA-based systolic wave detection algorithm was evaluated using 40 records after three heat stress simulations containing 5,071 heartbeats, with an overall SE of 99.89% and +P of 99.84%.[53] The TERMA-based systolic detector slightly outperformed existing algorithms, such as those of Billauer[83] (SE of 99.88% and +P of 98.69%), Li[117] (SE of 97.9% and +P of 99.93%), and Zong[118] (SE of 99.69% and +P of 99.71%).

- For a and b wave detection in APG signals: the TERMA-based a wave detection algorithm demonstrated an overall SE of 99.78% and a +P of 100% over signals that suffer from: (i) non-stationary effects; (ii) irregular heartbeats; and (iii) low amplitude waves. In addition, the b detection algorithm (based on the detection of a waves) achieved an overall SE of 99.78% and +P of 99.95%.[51] The TERMA-based a and b waves detector was not compared to other algorithms, as it is a new area of investigation and is considered a pioneering concept in the field of PPG signal analysis. However, the results are very promising as the scored accuracy over heat-stressed PPG signals is >98%.

- For c, d and e wave detection in APG signals: the performance of the TERMA-based c, d and e wave detector was tested on 27 PPG records collected during rest and after 2 hours of exercise, resulting in an SE of 97.39% and a +P of 99.82%.[52] The TERMA-based c, d and e wave detector was not compared to other algorithms, as it is a new area of investigation, and the work is a pioneering concept in the field of PPG signal analysis. However, the results are very promising, as the scored accuracy over heat-stressed PPG signals is >97%.

Given the simplicity and the fact that memory and CPU power are not a huge concern nowadays, the proposed TERMA-based algorithm presents a clear advantage over the previously reported algorithms in terms of detection performance over large datasets and different application problems.

7.6 DISCUSSION

Application of the TERMA-based detectors has been demonstrated in Section 7.5. It is now necessary to further elaborate on the implementation of TERMA-based detectors. It is worth noting that TERMA is simple and clearly laid out in comparison to other detectors published in the literature. For example, well-known algorithms demand more implementation steps[119] and resampling of the biosignals before processing; for example, the Pan–Tompkins algorithm[110] requires a resampling step for any ECG signal not sampled at 200 Hz. Its filters are designed for 200 Hz, so performance will be degraded at other sampling frequencies.

Furthermore, TERMA-based detectors are amplitude-independent, while well-known detectors, such as the Pan–Tompkins algorithm, are amplitude-dependent. Moreover, TERMA-based detectors use an efficient dynamic thresholding, while other algorithms, such as the Pan–Tompkins algorithm, have a complicated thresholding step to adjust the threshold. The TERMA-based algorithm does not need to change its threshold based on previous segments.

7.6.1 Frequency Band Choice

In the literature, most of the researchers developed detection algorithms and determined the frequency bands experimentally without justifying their choice. For example, researchers have used 5–15 Hz,[120] 5–11 Hz,[110] 4–13.5 Hz,[117] 4.1–33.1 Hz,[121] 9–30 Hz,[122] and 2.2–33.3 Hz[123] as the optimal frequency band for the detection of QRS complexes in ECG signals. However, the proposed TERMA method extracts the optimal frequency band during the training stage through a rigorous brute force optimization.

The choice of frequency band plays a major role in reducing the amount of noise in the processed signals. However, determining a reasonable estimate for the frequency band can be easily carried out on a part of the sample size using the power spectrum of the investigated event,[113] which step is relatively easier compared to determining the window sizes in the TERMA method.

The bandpass filter consists of two filters, the low-pass filter and the high-pass filter. The low-pass filter is used to remove high frequency noise, and the high-pass filter is used to remove low frequency noise. Usually, a Butterworth filter is used due to its simplicity and is characterized by a magnitude response that is maximally flat in the passband and is monotonic overall. MATLAB provides low-pass and high-pass filters with the simple command $butter(m, f, 'low')$ and $butter(m, f, 'high')$, respectively, where m is the filter order and f is the normalized cut-off frequency. The purpose of this step is to retain the characteristics of the main events within the processed signal, remove the undesired noise, and make the main events more salient.

7.6.2 Window Size Choice

After noise removal, the TERMA window sizes need to be determined. The two window sizes reflect the event duration and the event repetition period (cyclic duration), which is an individual characteristic that depends on the heart rate and abnormalities and, thus, is hard to predict. It is common that researchers determine the window size of a moving average without a proper justification or reasoning; for example, Pan and Tompkins[113] used one moving average to demarcate the QRS complexes in ECG signals with a window size of 150 ms. However, the proposed TERMA method overcomes unjustified window sizes and offers two event-related window sizes for the two moving averages. Therefore, the TERMA window sizes depend on the expected duration of the investigated event and the repetition period of this event. These window sizes can be adjusted via a predefined dataset, or can be optimized over a representative sample during the training phase, as discussed above.

In the TERMA method, the use of two moving averages does not always generate blocks of interest. When the two moving averages are able to generate blocks of interest, this is referred to as "coupled" moving averages. To understand and generalize the coupling process between the two moving averages in TERMA, the W_2/W_1 ratio needs to be examined. The coupling between the window sizes of the moving averages over different biomedical signals is investigated, as shown in Figure 7.4.

To assess the coupling between window sizes, the performance of the created TERMA detectors based on the generated blocks of interest is explored. The performance of TERMA detectors in terms of overall accuracy in detecting a particular event is split into two categories: coupling and non-coupling. The coupling category is when the two moving averages were able to generate blocks of interest and achieved an accuracy of >50%, while the non-coupling category is when the two moving averages were unable to generate blocks of interest and achieved an accuracy that is not-a-number (NaN).

Figure 7.4 demonstrates the effectiveness of the coupling process. For detecting the systolic and a waves in APG signals, the dominant ratio that is able to generate blocks of interest ($W_2/W_1 = 6$) is the optimal ratio, as shown in Figure 7.4i,ii, but TERMA did not fail over the investigated ratios during the training phase, as shown in Figure 7.4iii,iv. When detecting c, d, and e waves, the optimal coupling ratio is ($W_2/W_1 = 3$), as shown in Figure 7.4v, while the most non-coupling ratio is ($W_2/W_1 = 2$), as shown Figure 7.4vi. These results show that the optimal coupling for TERMA can be achieved using the following inequality:

$$(8 \times W_1) \geq W_2 \geq (2 \times W_1)$$

(7.7)

where the lower bound is ($2 \times W_1$) and the higher bound is ($8 \times W_1$).

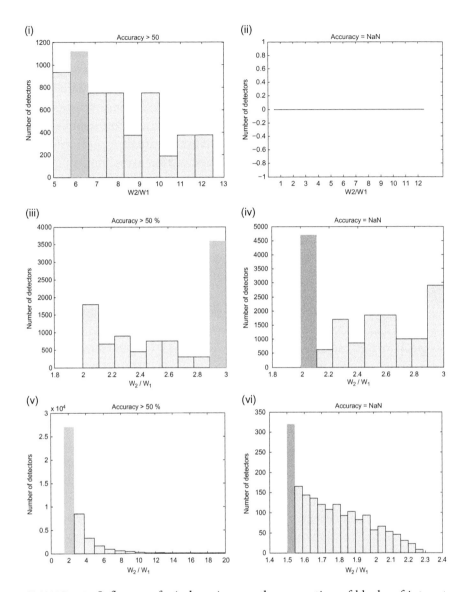

FIGURE 7.4 Influence of window sizes on the generation of blocks of interest based on overall accuracy.[109] *Notes*: (i,ii) systolic waves detection; (iii,iv) *a* waves detection; (v,vi) *c*, *d*, and *e* wave detection. The lefthand column represents the coupling between the two moving averages by scoring >50% accuracy, while the righthand column represents non-coupling. The coupling is referred to as accuracy >50%, while non-coupling is referred to as not-a-number (NaN). The light blue bar represents the most dominant W_2/W_1 ratio in coupling, while the purple bar represents the most dominant W_2/W_1 in non-coupling.

As can be seen in Figure 7.4, if the W_2 is not well-defined with respect to W_1, the detector fails to detect any events. The TERMA testing results, discussed above, are promising for handling the non-stationary effects, low SNR, left bundle branch block, right bundle branch block, premature ventricular contraction, premature atrial and fast heart rate over different biomedical signals. As it is a new concept, there is a need to publish the current results and let the scientific community evaluate its performance on their studies with different types of noise and abnormalities.

7.6.3 Offset β Choice

The offsetting step has been used in the literature as the last stage for most of the event detection algorithms.[124–130] The performance of the offsetting approach will be affected by low SNR signals.[110,131] Usually, the offset is a fixed value and is experimentally defined.[110,132,133] The offsets of these algorithms have been selected based on estimations, which, in turn, had an impact on the algorithms' performance. Because the offsetting approach is simple (just an IF-THEN-ELSE statement), researchers used it as a computationally efficient approach to improve accuracy.[110,132,133] The use of a fixed offset to detect a particular event is efficient for stationary biomedical signals with a normal beat morphology. Due to severe baseline drifting and the movement of patients, the waveforms of the collected biomedical signal may vary drastically from one heartbeat to the next. Therefore, the probability of missing events is high.

The TERMA offset will slightly shift the output of the second moving average up with the longer window size when applied as a threshold to the first moving average with the smaller window size. The use of a fixed offset to detect particular events, such as systolic waves, is simple and efficient for stationary PPG signals with normal beat morphology. Due to severe baseline drifting and the movement of patients, the PPG signal waveform may vary drastically from one heartbeat to the next. Therefore, the probability of missing systolic waves is high. With signal-dependent offsetting (as a percentage of the signal amplitude), the probability of missing events, such systolic and a waves, decreases.

To assess the impact of the offset on the coupling process, the performance of TERMA detectors is assessed in terms of overall accuracy (>50% and =NaN). Interestingly, the offsetting does not affect the coupling process, as shown Figure 7.5i,iii,v. In other words, the change in β does not affect the generation of blocks of interest but, rather, improves the overall accuracy of detection, especially when the processed signal is relatively noisy. Moreover, the offsetting step in the case of detecting systolic and *a* waves caused no non-coupling case, as shown in Figure 7.5(ii), (iv).

7.6.4 Battery-Driven Devices

Simplicity is particularly effective when it comes to mobile and battery-driven device computation. Simple analysis methods that achieve a high level of accuracy in event detection require less storage and power, and are more suitable for battery-driven devices.[119,134] It is important to mention that simplicity cannot be achieved at the expense of reliability. Simplicity goes hand-in-hand with reliability.[134]

A simple, yet efficient event detector is needed to provide a more accurate analysis for wearable devices, point-of-care devices, fitness trackers, and smart watches, especially when performance is compared to more complex machine learning solutions.[58,84] Event detection algorithms have been published in the literature[55,120] and compared based on numerical efficiency. It was concluded that the better the numerical efficiency, the faster the algorithm, and vice versa. In other words, the faster the algorithm, the more suitable it is for battery-driven devices.

In the conclusion of Elgendi et al. (2014),[119] the researchers recommended implementing moving averages for battery-driven devices, as they are highly numerically efficient. The implementation of one moving average to detect for simplicity has been discussed in Christov et al. (2004)[135] and Chiarugi et al.(2007).[136] However, the thresholding phase of these one moving average algorithms was complicated and increased the computational complexity.[119] On the other hand, the TERMA detector was more efficient and faster than the one moving average algorithms.[55]

It is intuitive to think that the use of one moving average is better than using two moving averages, especially for implementation on battery-driven devices. The problem with this approach is the decision-making steps required to detect the event. For example, the one moving average algorithm utilizes a fixed window size that is determined empirically, and thresholds depend on the accuracy of the heart rate determined in the previous segment.[55,110] However, the TERMA detector does not need to work with a fixed window size; in fact, TERMA processes the whole recording at once. Moreover, TERMA does not need to check the past segments or the previous detection rate.[55] TERMA is advantageous because it uses the second moving average as a threshold to the first moving average, without the need for any complicated thresholding. Therefore, TERMA is promising for a battery-driven device compared to other algorithms.

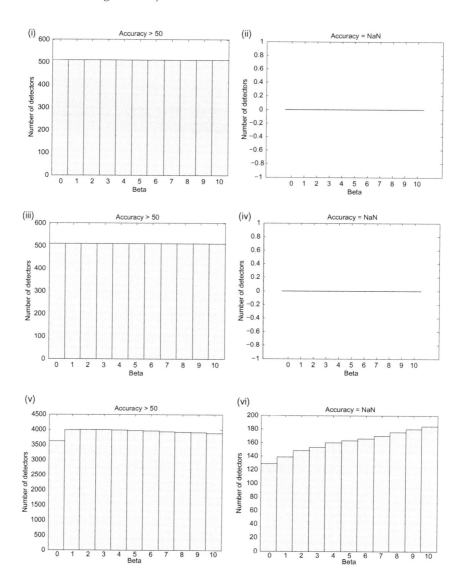

FIGURE 7.5   Influence of the offset (β) on the generation of blocks of interest on overall accuracy.[109] *Notes:* (i,ii) systolic wave detection; (iii,iv) *a* wave detection. The impact of the β value on the coupling by scoring >50% accuracy is represented in the lefthand column, while the non-coupling is represented in the righthand column. The coupling is referred to with accuracy >50%, while non-coupling is referred to as not-a-number (NaN).

7.6.5 Optimization Step

The multiple objective optimization step is time-consuming and is not computationally efficient. Perhaps the calculations of the optimization step could take place with the use of high-performance computers. However, the implementation of the optimization step is essential and only needs to be carried out once to find the optimal Pareto solution. Once the optimal solution is determined, the TERMA algorithm sets the optimal value as the fixed value for each parameter, and this can be implemented on battery-driven devices with low computational power. Notably, machine learning algorithms usually require a high degree of computational power for both the optimization step and the algorithm implementation step which can pose a challenge to battery-driven devices.

The TERMA prior knowledge step is important for the practical understanding of the expected characteristics of the events and noise within the signal. During optimization, the relationship between the processed signal, added noise, existing events, and TERMA parameters is considered. In the test phase, the optimal combination of parameters obtained from the optimization step will be used without any further adjustments. It is important to include a wide variety of waveform and noise to obtain the optimal combination that suits most cases.

In order to calculate the accuracy, the training signal needs to annotate the events that need to be extracted. We also need a correct criterion for event extraction, such as the difference between the time extracted by the algorithm and the annotated event being less than 20 ms. The following shows how to calculate accuracy:

```
1    x1 = AnnotateEvents;   %AnnotateEvents is
         the annotated events
2    x2 = ExtractedEvents;  % ExtractedEvents is
         the events extracted by TERMA detector
3    MaxDifference = 0.02;  % The maximum
         difference in judging the accuracy of an
         event
4    TP_inAnnotate = [];
5    TP_inExtracted = [];
6    FN = [];
7    FP = [];
8    for i = 1:length(x1)
9            t = find(abs(x1(i) - x2)
             <MaxDifference);
10           if ~ isempty(t)
```

```
11                    TP_inAnnotate = [TP_inAnnotate, i];
12                    TP_inExtracted = [TP_inExtracted, t];
13             end
14         end
15         TP = length(TP_inAnnotate);
16         FN_events = x1;
17         FN_events(TP_inAnnotate) = [];
18         FN = length(FN_events);
19
20         FP_events = x2;
21         FP_events(TP_inExtracted) = [];
22         FP = length(FP_events);
23
24         SE = TP/(TP + FN);
25         PP = TP/(TP + FP); %+P
```

Optimizing all parameters will require a search methodology. Based on the application and computational resources available, a search method can be selected. Two search methods will be discussed below: exhaustive search, and gradient-based search.

7.6.5.1 Exhaustive Search

The exhaustive (brute force) search is a simple and effective method to obtain the optimal parameters. By a given vector of selectable values for each parameter, the exhaustive method can find the best combination of parameters. The number of iterations is the product of the number of selectable values for all parameters. The code is as follows:

```
1 n = 1;       % the iterations
2 for F1 = F1_vector      % F1_vector is the selectable
                          values of parameter F1
3     for F2 = F2_vector % F2_vector is the
              selectable values of parameter F2
4         for W1 = W1_vector % W1_vector is the
                  selectable values of parameter W1
5             for W2 = W2_vector % W2_vector is the
                      selectable values of parameter W2
6                 for beta = beta_vector % beta_
                          vector is the selectable
                          values of parameter beta
7                     ExtractedEvents = DETECTOR
                          (F1 , F2 ,W1,W2, beta ) ; %
                          extract events by TERMA
```

```
8                          [SE(n) ,PP(n) ] =
                           calculateAccuracy
                           (AnnotateEvents ,
                           ExtractedEvents ) ;
                           %calculate the accuracy SE
                           and +P
9                          n = n + 1;
10                         parameters (n, :) =
                           [F1,F2,W1,W2, beta];
11                end
12             end
13          end
14      end
15 end
16 Results = [SE (:), PP (:), parameters];
17 SortedResults = sortrows ( Results, 'descend');
18 BestParameters = SortedResults (1, 3: end);
```

7.6.5.2 Gradient-Based Search

The exhaustive search takes a considerable amount of time, and a better method is using the optimization toolbox in MATLAB. The function *fmincon* is a gradient-based method that is designed to work on problems where the objective and constraint functions are both continuous and have continuous first derivatives. It can find the minimum of a constrained nonlinear multivariable function. The following shows an example of using the *fmincon* function to obtain the optimal parameters.

First, we should create a constrained function using the following code:

```
1 function [c,ceq] = ConstrainedFun(para)
2 %Linear inequality constraints
3 c(1)=para(1) - para(2);      % F1 < F2
4 c(2)=para(3) - para(4);      % W1 < W2
5
6 %Linear equality constraints
7 ceq = [];
8 end
```

Then, we can calculate the optimal parameters:

```
1
2 objectiveFun = @(para)(1 - calculateAccuracy(Annota
      teEvents , DETECTOR(para(1),para(2),para(3),
```

```
      para(4),para(5))))      % creat objecttive
      function, the para is a parameter vector which
      is as the format [F1, F2,W1,W2, beta]
3
4
5  para0 = [f1_init, f2_init, w1_init, w2_init, beta_
         init]; % the initial point of parameters vector
6  lpara = [f1_min, f2_min,w1_min, w2_min, beta_min];
         % the lower bounds of parameters
7  upara = f1_max, f2_max, w1_max, w2_max,beta_max];
         % the upper bounds of parameters
8
9  options = optimoptions('fmincon','Algorithm',
      'interior-point','StepTolerance',1e-10,
      'ConstraintTolerance',1e-10);  % the setting of
      fmincon function
10
11 optimalPara = fmincon(objectiveFun,para0,[],[],[],
      [],lpara,upara,@ConstrainedFun,options);
```

In this example, we used the interior-point algorithm to calculate the optimal parameters. The *fmincon* function is fast, but the results are likely to be different from the results obtained by the exhaustive search method. One reason is that the possible values of the parameters in the *fmincon* method are continuous, and the exhaustive method is discrete. Another reason is that the result obtained by *fmincon* may be a local minimum. Limited by resources and computing power, we would consider this to be a fast and effective solution.

7.6.5.3 Parallel Execution

Calculating optimal parameters takes a long time. We could also try the parallel method, such as using parallel for loop *parfor*. Note that, unlike a traditional for-loop, iterations are not executed in a guaranteed order. You cannot call scripts directly in a parfor-loop.

7.7 SIGNIFICANCE OF TERMA

We saw how TERMA-based detectors succeeded in detecting events, such as QRS, T, systolic, *a*, *b*, *c*, *d*, *e*, S1, and S2 in different biomedical signals. These biomedical signals were collected using different biosensors with different sampling frequencies in noisy environments. The databases used in the evaluation of TERMA contain signals suffering from: (i) non-stationary effects; (ii) low SNR; (iii) PACs; (iv) PVCs; (v) LBBBs; (vi) RBBBs; (vii) PAH; (viii) heat stress. Based on a review of the current literature, TERMA is the only framework that can be applied to different applications with great success.

The TERMA framework is not only reliable, but also numerically efficient and intuitive. It is easy to track the detection rate and to improve accuracy by adjusting five variables. As discussed above, the window sizes (W_1 and W_2) play a major role in detecting main events in biomedical signals. In other words, setting the values of W_1 and W_2 will enable fast analysis of the process. Adjusting the window sizes provides detailed information of the dominant events in terms of morphology and duration.

Results from this chapter lend more insight into implementing the block of interest generation step, by defining the relative values between W_1 and W_2 to be $[(8 \times W_1) \geq W_2 \geq (2 \times W_1)]$. These boundaries can be referred to as the "TERMA rate"; it defines the limits of the lower boundary and higher boundary of successful coupling between two moving averages ($[(8 \times W_1) \geq W_2 \geq (2 \times W_1)]$). This is similar to finding the boundaries for signal sampling (Nyquist rate $[f_s > 2f_{max}]$). To clarify the analogy, if we sample a signal at, or above, the Nyquist rate, we can reconstruct the signal. Similarly with the TERMA rate: if the second window size is larger than double the first window size and less than the octuple of the first window size, we can generate blocks of interest and detect main events. Moreover, the TERMA rate can be used to improve the recently published eventogram, a visualization tool that depends on two moving averages generating blocks of interest.[58]

The significance of the TERMA framework comes from its generic nature for the detection of patterns in any quasi-periodic signal. The TERMA framework consists of six independent steps, which can be perceived as LEGO building bricks; each one of these steps can be modified independently based on the detection problem. Thus, TERMA is flexible, universal, and can be applied to any periodic or quasi-periodic signals to achieve high accuracy in detecting dominant events within the processed signal. In other words, TERMA is a generic framework that enables researchers to change the filter type, filter order, and moving average type based on their application.

Exploring these findings across different types of periodic and quasi-periodic signals that have similar morphologies and characteristics will improve generalization across the entire signal analysis discipline.

Examples of similar morphologies and characteristics are the climatic time series[137] (which resembles noisy ECG signals), the plant electrical signal[138] (which resembles PPG signals), the optical signal[139] (which resembles PPG signals), the geophysical signal[140] (which resembles the NASDAQ Stock Market signal), the astrophysical signal[141] (which resembles noisy ECG signals), the geophysics signal[142] (which resembles noisy heart sounds), and the acoustic and vibration signal[143] (which resembles noisy heart sounds).

7.8 SUMMARY

Event detection in biomedical signals is an important step before analyzing the corresponding waveform in greater detail. A new economics-inspired approach for detecting events in biomedical signals is presented. The new algorithm is referred to as TERMA, and its functionality depends mainly on two moving averages similar to those used in economics to examine gross domestic product, employment, or other macroeconomic time series. The existence of prior knowledge about the examined waveforms within the biomedical signals will facilitate the adjustment of the window sizes of the two moving averages. Applying the optimization step provides the optimal values of the TERMA, which is recommended for higher detection accuracy. Once the optimal values of TERMA are determined, no further tuning is needed. Consequently, the validation of the same detector using another dataset without any later parameter tuning can help to obtain more reliable performance results. The performance of the TERMA-based detector is promising. It has been tested on different databases that contain unusual noise and different waveform morphologies. In the literature, it is common to find several algorithms to detect a particular event in a particular biomedical signal. The power of the TERMA-based detector is that it is a generic framework that can be applied to detect different types of events in different biomedical and quasi-periodic signals.

CHAPTER 8

Feature Selection

This chapter discusses the methods that can be used to select the most informative features from a PPG signal and its derivatives. Feature selection is an essential step as it: 1) finds out the optimal feature(s), 2) simplifies the predictive models, 3) reduces the training and testing times, 4) overcomes the curse of dimensionality, 5) reduces overfitting, and 6) reduces the amount of memory needed.

8.1 LEARNING OBJECTIVES

The learning objectives of this chapter are to:

- Learn about different methods for distinguishing two different data

- Develop an understanding of the separability tests

- Gain insight into how to rank key features by class separability criteria

8.2 FEATURE NORMALIZATION

Feature normalization is an important step for feature selection and for distinguishing objectively among different features. Without applying this step, the range of each feature will vary dramatically and the comparison cannot be carried out in a standardized manner. Moreover, without applying the normalization step, features with large values will have a stronger impact on decision-making during the comparison and classification processes. The feature normalization step provides an equal opportunity for each feature to show the influence it can bring to bear without the need to know the original ranges. Two common normalization techniques will be discussed: linear normalization, and nonlinear normalization.

8.2.1 Linear Normalization

The first normalization technique concerns normalizing the features to a zero mean and unit variance. If we have a feature set extracted from a PPG signal (or any derivative of a PPG) and saved in variable x, the normalization is carried out by subtracting the mean and dividing by the standard deviation, as follows:

$$\hat{x}_i = \left(x_i - \mu_x \right)/\sigma, \tag{8.1}$$

where $\hat{x}_i$ is the normalized feature while $\hat{\mu}_x$ and σ are, respectively, the empirical estimates of the mean and standard deviation of x_i; i refers to each feature of the feature set extracted from a PPG, VPG, or APG signal.

If we have a feature set in a variable called *Features*, and each row represents a feature vector, then we can write this in MATLAB as follows:

```
1 >> for i=1:FeatSet
2 >>     meanOfFeature=mean(Features(i,:));
3 >>     stdOfFeature=std(Features(i,:));
4 >>     NormalizedFeatures(i,:)=((Features(i,:)-
   meanOfFeature)/stdOfFeature);
5 >> end
```

The output of the MATLAB code is the variable *NormalizedFeatures*, which contains all features after the normalization step.

We can also obtain a feature set with values that ranges between 0 and 1, or between –1 and 1, using equation 8.2:

$$\hat{x}_i = MinVal + \left(\left(MaxVal - MinVal \right) * \left(x_i - x_{\min} \right) \right)/\left(x_{i\max} - x_{i\min} \right), \tag{8.2}$$

where $\hat{x}_i$ is the normalized feature. The maximum and minimum values of x_i are, respectively, x_{imax} and x_{imin}, and *MaxVal* and *MinVal* are, respectively, the desired lowest and highest values: i refers to each feature of the feature set extracted from a PPG, VPG, or APG signal. The MATLAB code for this normalization can be implemented as follows:

```
1 >>     MinVal = 0;     % desired minimum value
2 >>     MaxVal = 1;     % desired maximum value
```

```
3 >> for i=1:numFeat
4 >>       theMin=min(Features(i,:));
5 >>       theMax=max(Features(i,:));
6 >>       NormalizedFeatures(i,:)=MinVal+((MaxVal-
     MinVal)*(Features(i,:)-theMin))/(theMax-theMin);
7 >> end
```

8.2.2 Nonlinear Normalization

Another way to normalize features can be based on the amplitude range and uses a nonlinear method. For example, if the features are not evenly distributed around the mean, we can use the sigmoid function to generate a normalized feature set with values ranging between 0 and 1, as follows:

$$\hat{x}_i = 1/1 + \exp^{(1/r)*x_i}, \tag{8.3}$$

where $\hat{x}_i$ is the normalized feature. The r is determined by the user and, typically, the value of 0.5 is used; i refers to each feature of the feature set extracted from a PPG, VPG, or APG signal. Note that the sigmoid function returns values either from 0 to 1, or from –1 to 1, depending on the convention followed. The MATLAB code for sigmoid normalization can be implemented as follows:

```
1 >> r = 0.5;
2 >> Features=(1.0/r)* Features;
3 >> NormalizedFeatures=1.0 ./(1.0+exp(-Features));
```

8.3 CRITERIA FOR SELECTION AND EVALUATION

8.3.1 Independent Student's t-test

One of the first statistical tests to apply to your PPG feature set is the independent t-test. The type of the study determines the test independence; for example, the independent t-test is suitable for cross-sectional studies where researchers conduct several observations of different subjects. In other words, the independent t-test is used when the subject is not measured more than once.

If we extract features from PPG signals collected from different hypertensive and normotensive subjects, what we can do is examine the power of the separability of the extracted feature using a t-test. Note that, in this scenario, one PPG measurement per subject is used. The main idea here is that the t-test will examine whether the mean of the features collected from the independent hypertensive and normotensive subjects varies significantly across the range of subjects. In this case, we have two classes and the t-test assumes that the features set in both classes are normally distributed.

The classical way of writing the hypothesis is as follows:

1. H0: The mean of the independent feature vector in each class is equal.

2. H1: The mean of the independent feature vectors in each class is different.

The null hypothesis (H0) says that the means of the two independent feature vectors for the tested classes do not differ significantly from each other. The alternative hypothesis (H1) says that the values in the two classes differ significantly. If, after applying the t-test, the null hypothesis is true, then the the feature will not be selected, as it is not informative. If the alternative hypothesis is true, then the feature will be selected. The MATLAB function for the two-sample t-test is:

```
1 >> h = ttest2(x,y) % returns a test decision for the
     null hypothesis that the data in vectors x and y
     comes from independent random samples from normal
     distributions with equal means and equal but
     unknown variances, using the two-sample t-test .
     The alternative hypothesis is that the data in x
     and y comes from populations with unequal means.
     The result h is 1 if the test rejects the null
     hypothesis at the 5% significance level, and 0
     otherwise.
```

Regarding the application of the independent t-test:

1. Normality check: before running the t-test, it is recommended that the feature vectors are investigate to determine whether they are normally distributed. In MATLAB, there are two functions that can be used: the one-sample Kolmogorov-Smirnov test (*kstest*) and the Lilliefors test (*lillietest*):

```
1 >> h = kstest(x) % returns a test decision for
     the null hypothesis that the data in vector x
     comes from a standardnormal distribution,
```

against the alternative that it does not come from such a distribution, using the one-sample Kolmogorov–Smirnov test. The result h is 1 if the test rejects the null hypothesis at the 5% significance level, or 0 otherwise.

```
1 >> h = lillietest(x) % returns a test decision
     for the null hypothesis that the data in
     vector x comes from a distribution in the
     normal family, against the alternative that
     it does not come from such a distribution,
     using a Lilliefors test. The result is 1 if
     the test rejects the null hypothesis at the
     5% significance level, and 0 otherwise.
```

2. If the feature vectors are not normally distributed, then the Wilcoxon test can be used as follows:

```
1 p = ranksum(x,y) % returns the p-value of a
     two-sided Wilcoxon rank sum test: ranksum
     tests the null hypothesis that data in x and
     y are samples from continuous distributions
     with equal medians, against the alternative
     that they are not. The test assumes that the
     two samples are independent. x and y can
     have different lengths.
```

Note that the rank-sum test is equivalent to a Mann-Whitney U-test.

3. If you are not sure whether the sample size of your study is sufficiently large, use the MATLAB function *sampsizepwr*. This function computes the sample size, power, or alternative parameter value for the hypothesis test. In other words, you can compute the sample size required to achieve a certain power as follows:

```
1 >> nout = sampsizepwr(testtype,p0,p1) % returns
     the sample size, nout, required for a two-
     sided test of the type specified by testtype
     to have a power (probability of rejecting the
     null hypothesis when the alternative
     hypothesis is true) of 0.90 when the
     significance level (probability of rejecting
     the null hypothesis when the null hypothesis
```

is true) is 0.05. p0 specifies parameter
values under the null hypothesis. p1
specifies the value, or an array of values,
of the single parameter being tested under
the alternative hypothesis.

8.3.2 Dependent Samples (Paired) t-test

This section relates to a longitudinal study where PPG signals are measured from the same subjects over a period of time, sometimes lasting many years. For a sample of subjects with repeated measures or matched pairs of data points, the hypothesis can be written as follows:

- Null Hypothesis (H_0): the hypothesis is that there is no difference between pairs of data: ($H_0 : \mu_0 = 0$).

- Alternate Hypothesis (H_a): the hypothesis is that there is a significant difference between the pairs of data.

The null hypothesis (H0) says that the means of the two dependent feature vectors for the tested classes do not differ significantly from each other. In other words, when the subjects were measured again there was no significant difference. The alternative hypothesis (H1) says that there is a a a significant difference. In this case, if after applying the t-test the null hypothesis is shown to be true, then the the feature will not be selected, as it is not informative. If the alternative hypothesis is true, then the feature will be selected. The MATLAB function for the paired two-sample t-test is:

```
1 >> h = ttest(x,y) % returns a test decision for the
      null hypothesis that the data in x y comes from a
      normal distribution with mean equal to zero and
      unknown variance, using the paired-sample t-test.
```

8.3.3 Receiver Operating Characteristic Curve

The receiver operating characteristic (ROC) curve measures the separability power of a certain feature. The concept of ROC relies on the amount of overlap between the probability distribution functions representing the feature vectors in two classes. The ideal overlap occurs when two distributions are identical and, therefore, the value of the area under the receiver operating curve (AUC) equals zero. The ideal separation between the two distributions leads to $AUC = 1$, while $AUC = 0.5$ refers to a random guess for the separation power and is not desired.

The ROC visualizes the diagnostic ability of each tested feature using a particular separability test over its range of possible discrimination thresholds. Commonly, the label of the x axis refers to the false positive rate (FPR), and the y axis refers to the true positive rate (TPR).

```
1  [tpr,fpr,thresholds] = roc(irisTargets,irisOutputs) %
      The receiver operating characteristic is a
      metric used to check the quality of
      classifiers. For each class of a classifier ,
      roc applies threshold values across the
      interval [0 ,1] to outputs. For each threshold,
      two values are calculated: the True Positive
      Ratio (TPR) and the False Positive Ratio
      (FPR). For a particular class i , TPR is the
      number of outputs whose actual and predicted
      class is class i, divided by the number of
      outputs whose predicted class is class i. FPR
      is the number of outputs the actual class of
      which is not class i, but the predicted class
      is class i, divided by the number of outputs
      the predicted class of which is not class i.
```

The results of this MATLAB function with *plotroc* can be presented graphically as follows:

```
1  >> N1=100;  % number of normotensive subjects
2  >> N2=40;   % number of hypertensive subjects
3
4  >> mu1=3;
5  >> mu2=0;
6
7  >> var1=1;
8  >> var2=1;
9
10 >> Normtensive=mu1+var1*randn(1,N1);
11 >> Hypertensive=mu2+var2*randn(1,N2);
12
13 >> histogram(Normtensive)
14 >> hold on, histogram (Hypertensive)
15
16 >> xlabel('Feature value');
17 >> ylabel('Frequency');
```

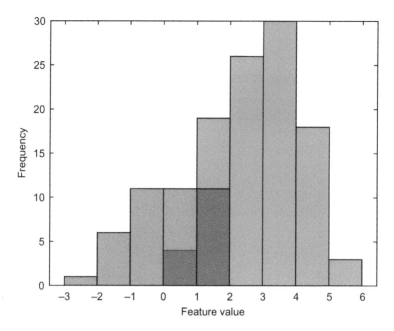

FIGURE 8.1 An example for probability distribution functions simulating a feature extracted from normotensive (blue) and hypertension (orange) subjects.

The previous MATLAB code generates Figure 8.1. As can be seen, there is overlap, the amount of which is difficult to determine visually. The ROC helps in showing the amount of overlap and separation. In Figure 8.2, the blue curve shows high values in terms of TPR and low values for FPR, given different thresholds. The ideal situation is to have a threshold that can achieve 100% TPR and 0% FPR, as shown in Figure 8.3 but this is not easy to achieve.

The *Confusion Matrix* is an alternative representation of the ROC result, as shown in Figure 8.4. It is also known as an error matrix. The difference between the confusion matrix and the ROC curve is that the confusion matrix visualizes the numerical results as a textual matrix (or table), rather than as a chart. Each row of the matrix represents the target values, and each column represents the output values.

8.3.4 Analysis of Variance (ANOVA)

In the case of multiple PPG features, there is a need to check whether all of them have the same separability power. ANOVA is able to compare between three or more feature means to determine whether the means are

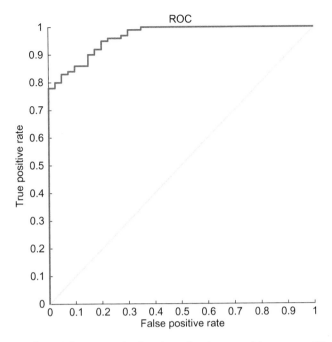

FIGURE 8.2   The ROC curve calculated on the data used to prepare Figure 8.1.

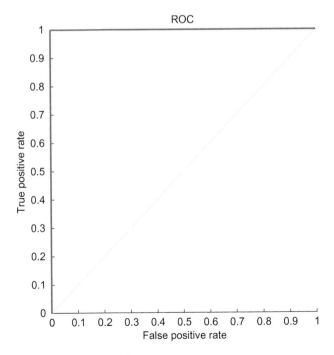

FIGURE 8.3   Optimal ROC result.

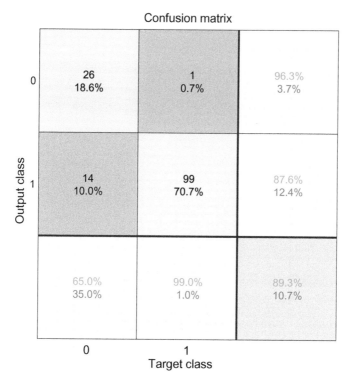

FIGURE 8.4 Confusion matrix obtained from the data used to prepare Figure 8.3.

equal. It is a more conservative generalization of multiple two-sample t-tests (less Type-I error). The main goal of applying ANOVA is to check whether features within a feature set are similar to each other (not useful), or vary (independent/useful features). All the parameters required for ANOVA are defined in Table 8.1. ANOVA uses the F-distribution and a corresponding F-test statistic to determine whether the group means are statistically significant as follows:

TABLE 8.1 Parameter calculation in ANOVA

	df	SS	MS	F
Between	$df_1 = k - 1$	$SSB = \sum n_j \left(\bar{X}_j - \bar{X} \right)^2$	$MSB = \dfrac{SSB}{df_1}$	$F = \dfrac{MSB}{MSE}$
Error	$df_2 = N - k$	$SSE = \sum \sum n_j \left(X - \bar{X}_j \right)^2$	$MSE = \dfrac{SSE}{df_2}$	
Total	$df_{total} = N - 1$	$SST = \sum \sum \left(X - \bar{X} \right)^2$		

- To determine if means are significant:
- Use F-table for desired α value (e.g., $\alpha = 0.05$)
- Find $F_{df1, df2}$
- If $F > F_{df1, df2}$ at desired α, observed differences between group means are significant!

The main MATLAB function that applies these mathematical steps, is *anova1* and defined as follows:

```
1  >> p = anova1(y) % returns the p-value for a balanced
      one-way ANOVA. It also displays the standard
      ANOVA table (tbl) and a box plot of the columns
      of y. anova1 tests the hypothesis that the
      samples in y are drawn from populations with the
      same mean against the alternative hypothesis that
      the population means are not all the same.
```

Here is an example of one way to implement ANOVA for PPG features. Let us assume that we need to test a certain feature extracted from a PPG signal (or its derivative) collected from normotensive, prehypertensive, and hypertensive subjects. If the result of ANOVA shows significance, then it is a good feature that can be used as a marker for blood pressure assessment. If not, then we ignore the tested feature and try a different one. The following MATLAB function shows how to apply ANOVA with simulated feature vectors.

```
1  >> u1=rand(100); % simulates feature values extracted
      from Normtensive subjects

2  >> u2=rand(100); % simulates feature values extracted
      from Prehypertensive subjects
3  >> u3=rand(100); % simulates feature values extracted
      from Hypertensive subjects
4
5  >> U = [u1 u2 u3];
6  >> anova1(U)
```

The result from ANOVA showed that there was no significance ($p = 0.4$), as reported in Figure 8.5, and that this feature (simulated feature) should therefore be ignored. The following MATLAB code

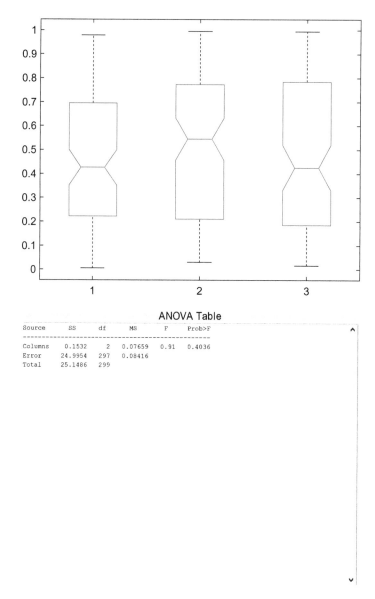

```
Source      SS        df    MS       F       Prob>F
-----------------------------------------------------
Columns    0.1532      2   0.07659   0.91    0.4036
Error     24.9954     297  0.08416
Total     25.1486     299
```

FIGURE 8.5   An example of ANOVA with no significance.

gives a simulated example for a feature set that can distinguish between three blood pressure levels:

```
1 >> M=meshgrid(1:100);
2 >> V=M(:,1:3);
3 >> W=U.*V;
4 >> anova1(W)
```

Figure 8.6 demonstrates significant differences between three blood pressure categories. In this case, the feature used is very informative and can be used as a biomarker for the blood pressure application.

Note that *anova*1(*y*) works on the same number of features extracted from the three blood pressure categories (i.e. a balanced design). On the

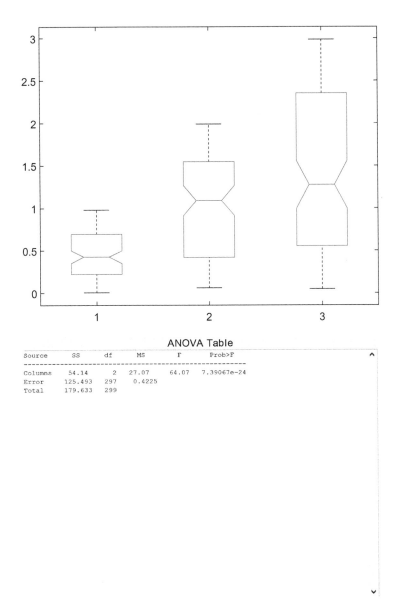

FIGURE 8.6 An example of ANOVA with significance.

other hand, *anova1(y,group)* can be used for features that have a different number of samples (i.e. an unbalanced design). Two other MATLAB functions are also appropriate: *anova2* and *anovan*.

8.3.5 Fisher's Measure

Fisher's measure (J_F) is commonly used to quantify the difference between two categories. The Fisher's measure takes into consideration two main statistical characteristics of a feature set: mean and variance. The ratio J_F is defined as follows:

$$J_f = \frac{(\mu_1 - \mu_2)}{\sigma_1^2 + \sigma_2^2},$$

(8.4)

where μ_1 and μ_2 are the statistical means, while σ_1^2 and σ_2^2 are the statistical variances of the two tested feature vectors. For example, if there is a feature extracted from normotensive and hypertensive subjects and there is a need to test the separability power of this feature, the ratio J_F can be used; the MATLAB code may be as follows:

```
1  >> m _ N = mean(F _ N); % F _ N is the PPG feature
      vector for normotensive subjects
2  >> m_H = mean(F_H); % F_H is the PPG feature vector
      for hypertensive subjects
3  >> v_N = var(F_N);
4  >> v_H = var(F_H);
5  >> J_F = (m_N - m_H)^2 / (v_N + v_H);
```

If the J_F value for feature 1 was higher than the J_F value for feature 2, then feature 1 is more informative than feature 2. The higher the value of J_F, the better the separation between the categories as the distance between the means causes relatively a larger value. Let us use the variables U and W discussed in the ANOVA section: we will find $J_F(W(:, 1), (W(:, 2))) = 0.731$ while $J_F(U(:, 1), (U(:, 2))) = 0.016$. As we knew beforehand that the W contains a feature that is more separable than U, the J_F obtained from W had a higher value than U, which confirms ANOVA's result.

In order to visualize how Fisher's can discriminate between two categories, two features are used, as shown in Figure 8.7. As can be seen, Fisher's ratio finds the projection of the feature vectors onto a line (or a plane, in the case of multiple features) that maximizes the distance between two categories.

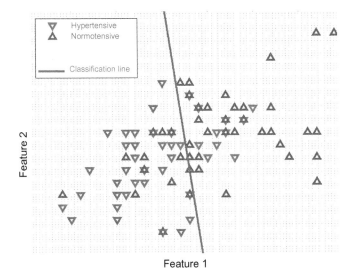

FIGURE 8.7   An example of Fisher's discriminant result.

8.3.6 Divergence Measure

The divergence measure (J_d) is similar to Fisher's measure but is used for multiple features. In other words, the term "feature matrix" will be used instead of the term "feature vector". It also assumes that the the feature vectors are normally distributed. The divergence distance J_d is defined as follows:

$$J_d = \frac{1}{2} trace\left\{ S_1^{-1} S_2 + S_2^{-1} S_1 - 2\mathbf{I} \right\} + \frac{1}{2}(m_1 - m_2)^T$$
$$\left(S_1^{-1} + S_2^{-1} \right)(m_1 - m_2), \tag{8.5}$$

where S_1 and S_2 are the covariance of feature matrices for two categories, $\mathbf{I}$ is the ll identity matrix, and μ_1 and μ_2 are the statistical means of the columns of the feature matrix. For simplicity, this measure is appropriate in situations where one feature is extracted from normotensive and hypertensive subjects, and there is a need to test the separability power of this feature.

The general MATLAB code for J_d measure can be written as follows:

```
1 >> Cov_N = cov(F_N); % F_N is the PPG feature matrix
     (multiple features) for normotensive subjects
```

```
2 >> Cov_H = cov(F_H); % F_H is the PPG feature
     matrix (multiple features) for hypertensive
     subjects
3 >> iCov_N = inv(Cov_N);
4 >> iCov_H = inv(Cov_H);
5 >> m_N = mean(F_N,1);
6 >> m_H = mean(F_H,1);
7 >> P1 = trace((iCov_N * Cov_H) + (iCov_H * Cov_N)
     - 2) /2;
8 >> x1 = (m_N - m_H);
9 >> x2 = (iCov_N + iCov_H);
10 >> P2 = (x1 * x2 * x1') / 2;
11 >> J_d = (P1 + P2) / size(F_N,2); % this step
     normalizes the output; however, it is not
     present in the equation above.
```

Similar to Fisher's measure output, if the J_d value for the combination of all features in matrix 1 was higher than the J_d value for all features in matrix 2, then matrix 1 is more informative than matrix 2. The higher the value of J_d, the combination of all features, the better the separation.

For simplicity, and to follow the previous examples in the Fisher's and ANOVA's sections, let us use variables U and W again to clarify the idea. We will find $J_d(W(:, 1), (W(:, 2))) = 5.81$, while $J_d(U(:, 1), (U(:, 2))) = 1.83$. As we knew beforehand that the W contains a feature that is more separable than U, the J_d obtained from W had a higher value than U, which confirms ANOVA's and Fisher's results. The only difference here is that the divergence measure improved the sensitivity of the separability power compared to that of Fisher's measure.

8.3.7 Bhattacharyya's Measure

Bhattacharyya's measure (J_B) is similar to the divergence measure. Also, it assumes normal distribution for all features. Bhattacharyya's measure J_B is defined as follows:

$$J_B = \frac{1}{8}(m_1 - m_2)^T \left(\frac{S_1 + S_2}{2}\right)^{-1}(m_1 - m_2) + \frac{1}{2}\ln\left(\frac{\|S_1 + S_2\|}{\sqrt{\|S_1\| + \|S_2\|}}\right), \quad (8.6)$$

where S_1 and S_2 are the covariance of feature matrices for two categories, $\|\ \|$ refers to the mathematical determinant of the matrix, and μ_1 and μ_2 are the statistical means of columns of the feature matrix. For simplicity, this measure is appropriate where one feature is extracted from normotensive and hypertensive subjects, and there is a need to test the separability power of this feature.

The general MATLAB code for J_B measure may be written as follows:

```
1 >> Cov_N = cov(F_N); % F_N is the PPG feature matrix
     (multiple features) for normotensive subjects
2 >> Cov_H = cov(F_H); % F_H is the PPG feature matrix
     (multiple features) for hypertensive subjects
3 >> m_N = mean(F_N,1);
4 >> m_H = mean(F_H,1);
5 >> P1 = (m_N - m_H);
6 >> P2 = inv((Cov_N + Cov_H)/2);
7 >> P3 = 0.5*log(det(Cov_N + Cov_H)/sqrt(det(Cov_N) +
     det(Cov_H)));
8 >> P4 = (P1' * P2 * P1)/8 + P3;
9 >> J_d = P4 / size(F_N,2); % this step normalizes
     the output; however, it is not present in the
     equation above.
```

Similar to the divergence measure output, if the J_B value for the combination of all features in matrix 1 was higher than the J_B value for all features in matrix 2, then matrix 1 is more informative than matrix 2. The higher the value of J_B, the combination of all features, the better the separation.

For simplicity, and to follow the examples in the previous sections, let us use variables U and W again to clarify the idea. We will find $J_B(W(:,1),(W(:,2))) = 0.414$, while $J_B(U(:,1),(U(:,2))) = 0.227$. As we knew beforehand that the W contains a feature that is more separable than U, the J_B obtained from W had a higher value than U, which confirms ANOVA's, Fisher's, and the divergence measure results. The only difference here is that the J_B measure did not improve the sensitivity of the separability power compared to that of the divergence measure.

8.3.8 Scatter Measure

The scatter measure (J_s) is one of the popular methods used to quantify the distances of the feature matrix in the feature space. It is similar to the divergence measure but more generic, and can be used over a large number of features simultaneously. Therefore, it is typically used for large feature sets. The scatter measure J_s is defined as follows:

$$J_s = trace\{S_w^{-1}S_m\}, \tag{8.7}$$

where S_w is the within covariance (summation of both covariance matrices), and S_m is the covariance of both feature matrices.

The general MATLAB code for the J_s measure can be written as follows:

```
1 >> Cov_N = cov(F_N,1); % F_N is the PPG feature matrix
     (multiple features) for normotensive subjects
2 >> Cov_H = cov(F_H,1); % F_H is the PPG feature matrix
     (multiple features) for hypertensive subjects
3 >> n1 = size(F_N,1);
4 >> n2 = size(F_H,1);
5 >> N = n1 + n2;
6 >> Sw = ( (n1/N)*Cov_N + (n2/N)*Cov_H );
7 >> c = [F_N;F_H];
8 >> Sm = cov(c,1);
9 >> J_s = trace( inv(Sw)*Sm ) / size(F_N,2);
```

Similar to the divergence measure output, if the J_s value for the combination of all features in matrix 1 was higher than the J_s value for all features in matrix 2, then matrix 1 is more informative than matrix 2. The higher the value of J_s, the combination of all features, the better the separation.

It is worth noting that the scatter measure can different forms, such as those shown in equations 8.8 and 8.9:

$$J_s = \frac{trace\{S_m\}}{trace\{S_w\}}. \tag{8.8}$$

$$J_s = \frac{\|S_m\|}{\|S_w\|}. \tag{8.9}$$

The first form in equation (8.7) will be used to explain the concept. For simplicity, and to follow the examples in the previous sections, let us use variables U and W again to clarify the idea. We will find $J_s(W(:,1),(W(:,2))) = 0.834$, while $J_s(U(:,1),(U(:,2))) = 0.458$. As we knew beforehand that the W contains a feature that is more separable than U, the J_s obtained from W had a higher value than U, which confirms the ANOVA's, Fisher's, the divergence, and Bhattacharyya's results. Note that the scatter measure improved the sensitivity of the separability power compared to that in Bhattacharyya's measure. In this example, the divergence measure was optimal in showing a significant difference for the same feature.

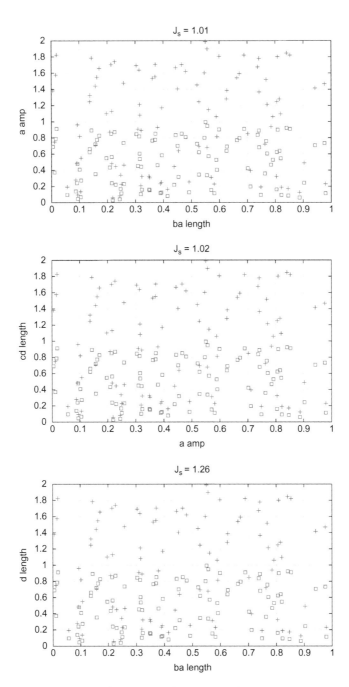

FIGURE 8.8 An example of a two APG features combination. *Notes*: It is clear that the combination of the two features in the bottom led to a better separation compared to the other two combinations. Note that the blue squares refer to normotensive subjects, while the red pluses refer to hypertensive subjects. Note that the higher the value of J_s, the better the separation. Note that not all features are normalized.

An example with two features is shown in Figure 8.8, after applying the scatter measure. The scatter value for feature1 and feature2 was J_s = 1.01, for feature2 and feature3 was J_s = 1.02, and feature1 and feature3 was J_s = 1.26. Based on these results, the combinaton of feature1 and feature3 is optimal, and more informative than the other combinations.

The question that can be raised at this point is what could we expect if we evaluated three features at once. In other words, is the combination of the three features likely to improve the scatter separability. After combining the three features, the scatter value scored was J_s = 1.27, as shown in Figure 8.9, which gives a better result than combining feature1 and feature3. Note that, if the value J_s = 1, this means that there is no separation between the two categories, given the number of features in hand.

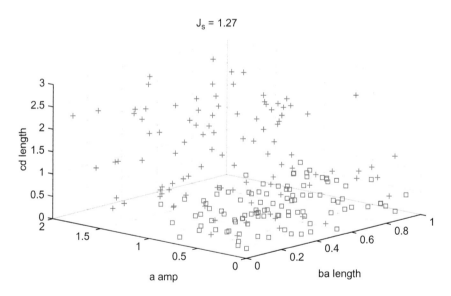

FIGURE 8.9   An example of a three APG features combination. *Notes*: It is clear that the combination of these three features provides a good separation between normotensive (blue squares) and hypertensive subjects (red pluses). Note that J_s = 1.27 and that two features are not normalized.

8.4 OPTIMAL FEATURE(S)

After discussing various measures to differentiate between features, it is important to discuss how to select the optimal feature or subset of features. The process of determining the optimal feature or subset of features is important, as it:

- simplifies models for easier interpretation;

- speeds up training with fewer features;

- reduces dimensionality with fewer features;

- reduces overfitting with fewer redundant features; and

- improves generalization and accuracy, as more features would make a diagnostic decision based on noise and would decrease overall accuracy.

The main goal here is to identify a subset of features that are relevant for constructing a diagnostic model, while excluding features that may be low-impact, or simply introduce noise and unnecessary complication into the desired model.

There are three main groups of feature selection methods: the statistical measure, the wrapper, and the embedded method.

- *Statistical Measure*:
  Searchings through the space of features evaluates a statistical criterion for each feature or subset of features. It applies a statistical measure to assign a scoring to each feature. The features are ranked in descending order according to their the score. Then, based on the ranking and the criterion, which is determined by the developer, the best feature or best two features will be retained. The statistical method is often used for finding the best feature after evaluating each feature independently. Examples of filters are the separability measures (discussed above), consistency measures, and correlation measures.

- *Wrapper*:
  Searching through the space of features evaluates a model for each subset. The wrapper method considers the selection of a set of

features as a search problem, where different combinations are prepared, evaluated, and compared to other combinations. A model is used to evaluate a combination of features and to assign a score based on the model's accuracy. An example of a wrapper method is the recursive feature elimination algorithm.

- *Embedded*:
  This feature selection method is embedded within the model, and is model-specific. Embedded methods learn which features best contribute to the accuracy of the model during the process of creating the model. The most common type of embedded feature selection method comprises regularization methods such as LASSO, Elastic Net, and Ridge Regression.

In all cases, feature selection is an optimization problem, as feature selection tries to find the optimal feature subset. The optimization is performed based on a criterion or objective function, while optimization tries either to maximize or to minimize the objective function.

8.4.1 Individual Feature Selection

One of the first steps to evaluate the available features individually is to use a statistical measure (t-test, Fisher's test, ROC, and so on). Then the features are ranked in descending order based on the separability measure; the first feature is the most informative feature, and the second most informative feature takes second place. A simple MATLAB code follows, and the method for finding the optimal feature is called an "exhaustive search":

```
1 >> x1 = F_N; % F _ N is the PPG feature matrix
     (multiple features) for normotensive subjects
2 >> x2 = F_H; % F _ H is the PPG feature matrix
     (multiple features) for hypertensive subjects
3
4 >> for i=1:size(x1,1)
5
6 >>      J = MEASURE(x1(i,:),x2(i,:); % J is the
     separability measure for any function MEASURE
7
8 >>      temp(i,1) = J;
9 >>      temp(i,2) = i;
```

```
10
11 >> end
12
13 >> [RankedFeatures,Index]=sortrows(temp,-1);
```

These simple steps are essential and comprise the main concept in all the feature selection methods. Also, note that the default for the MATLAB function *sortrows* is ranking in an ascending order. The defintion of *sortrows* is as follows:

```
1 B = sortrows( _ _ _ ,direction) % sorts the rows of
      A in the order specified by direction for any
      of the previous syntaxes.direction can be
      'ascend' ( default ) for ascending order, or
      'descend' for descending order. direction
      can also be a cell array whose elements
      are 'ascend' and 'descend' , where each
      element corresponds to a column on which
      sortrows operates. For example, sortrows
      (A,[4 6],{' ascend' 'descend'}) sorts the rows
      of A in ascending order based on the fourth
      column, then in descending order based on the
      sixth column to break ties.
```

After testing 100 features using the divergence measure, the output of this individual feature ranking using the separability measure *J* was as follows:

```
1 1.4860      31.0000 % the separabilty value is 1.48
      and feature #31 is the "ba area". In this case,
      this feature is the most informative feature.
2 1.2819      23.0000 % the separabilty value is 1.28
      and feature #23 is the "ab length". In this
      case, this feature is the second most
      informative feature.
3 1.1881      36.0000 % the separabilty value is 1.18
      and feature #36 is the "cd slope".
4 0.9867      48.0000 % the separabilty value is 0.98
      and feature #48 is the "b/a ratio".
5 0.9378      53.0000
6 .            .
7 .            .
```

```
 8  .                .
 9  .                .
10  0.0939        38.0000
11  0.0780        65.0000
12  0.0758        34.0000
13  0.0593        69.0000
14  0.0587        22.0000
15  0.0435         6.0000
16  0.0417        74.0000
17  0.0202        80.0000
18  0.0181        32.0000
19  0.0143        85.0000 % the separabilty value is 0.014
        and feature #85 is the "a amplitude". In this
        case, this feature is the least informative
        feature.
```

After testing the separability power for each feature individually and ranking the output, it can be seen that the feature "ba area" is the most informative feature for distinguishing between normotensive and hypertensive subjects. It also shows the least most informative feature—in this case, the "a amplitude". Note, this is not the ground truth; it is an example to show how the feature selection process is carried out. This result depends on many aspects, such as the sensor type, digitization, normalization, filtration, and so on.

Also note that the simple approach used above did not take into account the existing correlations among the features. Testing the cross-correlation coefficient over features can validate the overall finding. So, this is an extra step after obtaining the *RankedFeatures* from the individual feature test discussed above. The results after applying the cross-correlation are as follows:

```
1  31 % feature #31 is the "ba area". In this case,
       this featureis the most informative feature.

2  85 % feature #85 is the "a amplitude". In this
       case, this feature is the second most
       informative feature.
3  88
4  49
5  23
6  36
```

```
 7  .
 8  .
 9  .
10  .
11  57
12  78
13  19
14  21
15  2
16  93 % feature #93 is the "c amplitude". In this
       case, this feature is the second most
       informative feature.
```

As can be seen, feature #31 is consistent and this result confirms the result of the scatter measure; therefore, the conclusion is that feature #31 is the most informative feature.

8.5 SEARCH METHOD

Searching for the optimal feature(s) can be carried out using various logical steps. The search method tries to reach to the goal, which is finding the optimal feature or subset of features, as swiftly as possible with a minimal number of computational steps. There are two main search methods:

- Optimal search (numerically inefficient)
- Suboptimal search (numerically efficient)

8.5.1 Optimal Search

The optimal search is an exhaustive search (also called a brute force search), which is numerically inefficient. An example of an exhaustive search is discussed in sections 7.4.1 and 7.5.1 where all possible combinations will be "exhaustively" tested. In the optimal search, the separability measure is calculated for each combination of feature sets. For example, if the main goal is to determine the optimal combination of 4 features within a set of 100 features to detect hypertension, the MATLAB code for the exhaustive search may be done as follow:

```
1  >> x1 = F_N; % F_N is the PPG feature matrix
      (multiple features) for normotensive subjects
```

```
2 >> x2 = F_H; % F_H is the PPG feature matrix
     (multiple features) for hypertensive subjects
3 >> NumOfRequiredFeatures = 4; % this variable is
     determined by the user
4 >> NofFeatures = 100; % 100 features in hand
5 >> J_max=0;
6 >> Best_Features=[];
7
8 >> Combinations_all=nchoosek(1:NofFeatures,NumOfReq
     uiredFeatures);
9 >> noOfCombinations=size(cL,1);
10
11 >> for i=1:noOfCombinations
12 >>      J = MEASURE(x1(i,:),x2(i,:)); % J is
        the separability measure for any function
        MEASURE
13 >>      if J > J_max
14 >>          J_max = J;
15 >>          Best_Features = Combinations_all(i,:);
16 >>      end
17 >> end
```

Note the for loop going through all combinations. The J value is obtained by the *MEASURE* function, which is a separability measure. It is recommended that the codes described above on separability measures are converted to MATLAB functions to be able to call them up easily when needed. At the end of this search, the variable J_max will contain the highest separability value for all feature combinations, and the variable $Best_Features$ will contain the optimal feature combination.

To elaborate on the numerical complexity of the optimal search technique, let us assume we have a feature set of 100 features and that we are looking for the optimal two features: the number of iterations required is 4,950. The MATLAB code for calculating this example is:

```
1 >> factorial(100)/(factorial(100-2)*factorial(2))
```

The results of the optimal search on using U and W feature matrices, and divergence as a separability measure, used in the examples above, are:

```
1 Best_Features =
2
3       11        51
```

```
4
5
6  J_max =
7
8      1.3097e+08
```

According to the optimal search, the optimal two features are feature#11 and feature#51, with a large *J* value.

8.5.2 Suboptimal Search

It is clear that the exhaustive (brute force) search is the optimal search method with which to find the optimal feature set; however, it is computationally inefficient, especially if the feature set is large. There are faster (numerically efficient) search methods that can arrive at relatively optimal feature combinations, such as the sequential forward search, sequential backward search, sequential forward floating search, and sequential backward floating search.

Following our previous example in the optimal search, which has a feature set of 100 features from we are looking for the optimal two features, then the number of iterations required in the sequential forward search is 199 iterations. In other words, the sequential forward search improved the speed by 24 (from 4,950 iterations to 199 iterations). The MATLAB code for calculating this example is:

```
1
2  >> x1 = F_N; % F_N is the PPG feature matrix
        (multiple features) for normotensive subjects
3  >> x2 = F_H; % F_H is the PPG feature matrix
        (multiple features) for hypertensive subjects
4  >> NumOfRequiredFeatures = 2; % this variable is
        determined by the user
5  >> NofFeatures = 100; % 100 features in hand
6  >> J_max=0;
7  >> Best_Features=[];
8  >> counter=1;
9  >> while k <= NumOfRequiredFeatures
10 >>      maxJ=0;
11 >>      for i=1:NofFeatures
12 >>          if isempty(find(Best_Features==i))
13 >>              f = [Best_Features i];
14 >>          else continue;
15 >>          end
```

```
16 >>          J = MEASURE(x1(i,:),x2(i,:)); % J is the
    separability measure for any function MEASURE
17 >>          if J> maxJ
18 >>            J_max = J;
19 >>              Best_Features = f;
20 >>          end
21 >>       end
22 >>       counter = counter + 1;
23 >> end
```

It is clear that the sequential search depends on the first selected feature. In other words, the second feature will be selected based on the first selected feature. It is not usually the combination with the optimal first feature that improves the separability measure. The sequential search is a dependent search, while the exhaustive search is an independent search.

Let us check the results of the sequential search using the variable of the previous example (U and W feature matrices, and divergence as a separability measure). As expected, the results are not consistent with the optimal search and are as follows:

```
1 Best_Features =
2
3    29        31
4
5
6 J_max =
7
8       6.5104e+06
```

According to the sequential forward search, the optimal two features are feature #29 and feature #31, with a large J value. Note that the value of J from the sequential forward search is less than that of the optimal search. This means we have another good combination of two features, but not the optimal combination. However, given the resources and computational power, this is considered a swift and effective result. The same holds for the other suboptimal search methods.

8.6 SUMMARY

Domain knowledge regarding PPG features plays an important role in constructing the feature set. After extracting the features, normalization is required. If there are limited resources for your application, then finding one optimal feature can be valuable. Selecting a variety of features in combination can provide better results than one optimal feature. If the feature set is noisy and contains outliers, try to remove these before running any feature selection method. Comparing different feature selection measures can confirm the separability power of the each feature or the feature subset. The feature(s) with the highest degree of consistency in differentiation between the categories of interest will be those selected as the optimal feature(s). The search mechanism depends on computational resources. It is always recommended that feature combinations are checked using the optimal search, even if it is going to take more time.

CHAPTER 9

Identifying Adverse Events

This chapter discusses methods that can be used to classify various cases, usually abnormal PPG and normal PPG signals. Different classifiers, especially neural networks, have been used extensively in different applications. However, the abnormality classification in PPG signals is relatively new, and has attractive potential for the near future.

9.1 LEARNING OBJECTIVES

The learning objectives of this chapter are to:

- Develop an understanding of different types of classifier

- Learn how to adjust and test a classifier

- Gain insight into how to use MATLAB classification functions

9.2 MINIMUM DISTANCE CLASSIFIER

Usually, distance classifiers are among the first classifiers to be used for test and evaluation. They are simple to implement and rely on a concept that makes sense to scientists from a variety of backgrounds. The main concept behind a distance classifier is that if the feature being tested is near the mean of a class, then it belongs to that class. From a mathematical perspective, the feature F is assigned to the class C_1 if:

$$\sqrt{(F-m_1)^T V^{-1} (F-m_1)} < \sqrt{(F-m_2)^T V^{-1} (F-m_2)} \qquad (9.1)$$

where F is the feature matrix, m_1 and m_2 are the mean vectors, and V is the covariance matrix. Note that if $V = \mathbb{I}$, then the Mahalanobis distance becomes to the Euclidean distance. If the covariance matrix is diagonal, then the resulting distance measure is called a "standardized Euclidean distance", which is written as follows:

$$J = \sqrt{\sum_{i=1}^{N} \frac{(F-m_i)^2}{V_i}}. \qquad (9.2)$$

The MATLAB implementation of the Mahalanobis distance classification is as follows:

```
1   %%%%%%%%%%%%%%%%%%%%%%%%%%%%%%%%%%%%%%%%%%%%
2   % The following function performs
3   % the Mahalanobis distance classification
4
5   function [J] = maha_dist(m,V,Feature)
6   %%%%%%%%%%%%%%%%%%%%%%%%%%%%%%%%%%%%%%%%%%%%%%%%%%%%%
7   % Input variables:
8   %    m: mean vector of the normal distribution.
9   %    V: covariance matrix of the normal
            distribution. (Kx1)
10  %    Feature: the feature that will be tested.
11  %
12  % Output variable:
13  %    J: the class label
14  %
15  %%%%%%%%%%%%%%%%%%%%%%%%%%%%%%%%%%%%%%%%%%%%%%%%%%%%%
        %%
16
17  [1,K]=size(m);
18  [1,F]=size(Feature);
19
20  for i=1:F
21      for ii=1:K
22          J(j)=sqrt((Feature(:,i)-m(:,ii))'*inv(V)*
                (Feature(:,i)-m(:,ii)));
```

```
23     end
24     [num,J(i)]=min(J);
25 end
```

Having created the Mahalanobis distance function, it is appropriate to provide an example showing how to use the function and how to apply Mahalanobis decision theory. Let us extract two features from two PPG signals (one PPG signal collected from a hypertensive subject, the other from a normotensive subject), resulting in *Feature*1 = [1.12 0.11] and *Feature*2 = [0.7 0.33]:

```
1 % Extracting two features from two PPG signals
2 % The first signal is normotensive and the second
     signal is hypertensive
3 Feature1 = [1.12 0.11]';
4 Feature2 = [0.7 0.33]';
5 % let's means m1 and m2
6 m1 = [1.12 0.11]';
7 m2 = [0.7 0.33]';
8 % Coveriance matrix = identity matrix
9 V = eye(2);
10 % mean matrix
11 m = ([m1 m2]);
12 S = eye(2);
13 J1 = maha_dist(m,S,Feature1)
14 J2 = maha_dist(m,S,Feature2)
```

The results of the previous MATLAB code, which are the *J*1 and *J*e, are the following:

```
1 J1 =
2
3        1
4
5
6 J2 =
7
8        2
```

Note that $J1 = 1$, which refers to the feature vector *Feature*1 = [1.12 0.11] is classified as C_1. Similarly, $J2 = 2$, which refers to the feature vector *Feature*2 = [0.7 0.33] is classified as C_2. As anticipated, feature1 belongs to class 1 and feature2 belongs to class 2, which is what the distance classifier suggested.

9.3 BAYES CLASSIFIER

The Bayes classifier is a simple "probabilistic" classifier, developed based on Bayes' theorem. It is highly scalable, requiring a number of parameters linear in the number of variables (features/predictors) in a learning problem. This concept is crucial to understanding the core idea behind classification and pattern recognition. The Bayes classifer is easy to understand and to implement.

Let us start by clarifying the Bayes' decision theorem. The feature F is assigned to the class C_1 if:

$$P(C_1)p(F/C_1) > P(C_2)p(F/C_2) \qquad (9.3)$$

where $P(C_1)$ and $P(C_2)$ are, respectively, the a priori probabilities of class C_1 and C_2. The $p(C_1|F)$ and $p(C_2|F)$ are, respectively, the a posteriori probabilities of class C_1 and C_2 given the feature F. These are calculated using equations 9.4 and 9.5:

$$p(C_1|F) = \frac{1}{(2*\pi)^{K/2} |V^{0.5}|} \exp^{-0.5*(F-m_1)'V^{-1}(F-m_1)}, \qquad (9.4)$$

$$p(C_2|F) = \frac{1}{(2*\pi)^{K/2} |V^{0.5}|} \exp^{-0.5*(F-m_2)'V^{-1}(F-m_2)}. \qquad (9.5)$$

The MATLAB code that performs these steps is:

```
1  %%%%%%%%%%%%%%%%%%%%%%%%%%%%%%%%%%%%%%%%%%%%
2  % The following function calculates
3  % the probability density of normal distribution
4  function [p] = Gauss_posteriori (m,V, Feature)
5  %
6  % Input variables:
7  % m: mean vector of the normal distribution.
8  % V: covariance matrix of the normal distribution.
    (Kx1)
9  % Feature: the feature that will be tested .
10 %
11 % Output variable:
12 % p: the value of the normal distribution for the
    Feature variable.
13 %%%%%%%%%%%%%%%%%%%%%%%%%%%%%%%%%%%%%%%%%%%%
14
```

```
15  [K,Z]  =  size(m);
16  p_x  =  (1/((2*pi)^(K/2)*det(V)^0.5))*exp(-0.5*
        (Feature-m)'*inv(V)*(Feature-m));
```

Having created a function that has calculated the Gaussian probability density function, it is appropriate to provide an example showing how to use the function and how apply Bayes' decision theory. Let us extract two features from two PPG signals (one PPG signal collected from a hypertensive subject, the other from a normotensive subject), resulting in *Feature1* = [1.12 0.11] and *Feature2* = [0.7 0.33]:

```
 1  % Extracting two features from two PPG signals
 2  % The first signal is normotensive and the second
        signal is hypertensive
 3  Feature1 = [1.12 0.11]';
 4  Feature2 = [0.7 0.33]';
 5  % the probability of getting either classes is
        equal
 6  P1 = 0.5; % a priori probability of class 1
 7  P2 = 0.5; % a priori probability of class 2
 8  % to clarify the idea of using Gauss function
 9  % let 's means m1 and m2
10  m1 = [1.12 0.11]';
11  m2 = [0.7 0.33]';
12  % Coveriance matrix is identity matrix
13  V = eye(2);
14  % Calculating the posteriori probabilities
15  p1 = P1*Gauss_posteriori(m1,V,Feature1)
16  p2 = P2*Gauss_posteriori(m2,V,Feature1)
17  p3 = P1*Gauss_posteriori(m1,V,Feature2)
18  p4 = P2*Gauss_posteriori(m2,V,Feature2)
```

The results of the previous MATLAB code are the following:

```
 1  >>
 2
 3  p1 =
 4
 5       0.0796
 6
 7
 8  p2 =
 9
10       0.0711
```

```
11
12
13  p3 =
14
15        0.0711
16
17
18  p4 =
19
20        0.0796
```

It can be seen that $p1 > p2$, which refers to the feature vector *Feature1* = [1.12 0.11], is classified as C_1. Similarly, $p4 > p3$, which refers to the feature vector *Feature2* = [0.7 0.33], is classified C_2. This has occurred because the mean vectors *m1* and *m2* were set to be the exact vectors for *Feature1* and *Feature2*. Clearly, this is not the case in a real application; it is simply to show how the algorithm works. When developing a real application, the values of the *m* and *V* need to be determined before calling up the function. Note that, in this example, $P(C_1) = P(C_1) = 0.5$, which will vary according to the diagnostic problem.

9.4 COMPETITIVE NEURAL NETWORK

A competitive neural network is a form of unsupervised learning. There is no target here that has the same size as the input; rather, there are nodes competing to respond to the input data. Thus, the result of the competitive neural networks categorizes the input into a number of categories determined by the user.

In MATLAB, it is easy to construct a competitive learning neural network using the function *competlayer*. Note that the function *newc* is obsolete in R2010b NNET 7.0 and was last used in R2010a NNET 6.0.4: the recommended function is *competlayer*. Assume that we extracted two features from normotensive and hypertensive subjects, as was the case in the previous examples, and we need to categorize the two features in two classes:

```
1  >> Feature1 = [1.12  0.11];
2  >> Feature2 = [0.7  0.33];
```

Following extraction of the features from PPG signals, it is time to create a competitive neural network that has two outputs. Note that two outputs means two neurons, which means that two classes are expected. This step is implemented as follows:

```
1  >> net1 = competlayer(2);
```

Note that the data structure type of *net*1 is 'Network' in MATLAB. We can now view the weights for the connection from the first input to the first layer. The weights for a connection from an input to a layer are stored in net.IW. If the values are not yet set, these results will be empty, use the following:

```
1  >> net1 = net.IW{1};
```

You can view the bias values for the first layer as follows:

```
1  net.b{1}
```

The network structure contains the weights and biases of the neural network and can be accessed easily as follows:

```
1  >> net1.IW{1}
2
3  ans =
4
5     2 0    empty double matrix
6
7  >> net1.b{1}
8
9  ans =
10
11        5.4366
12        5.4366
```

Note that the weight values are not yet initialized, while the bias was set to [5.4366 5.4366]. After training the neural network, let us check the weights and biases. Here is how to train the neural network:

```
1  >> feature = [Feature1; Feature2];
2  >> net = train(net1,feature);
```

Let us check the weights and biases:

```
1  >> net.IW{1}
2
3  ans =
4
5        1.1200        0.7000
```

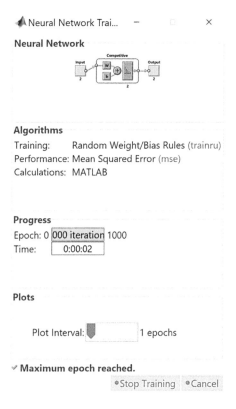

FIGURE 9.1   Interactive learning process using the MATLAB function *train*.
Note: Observe that the output layers contain two neurons to categorize the input
into two classes.

```
 6        0.1100        0.3300
 7
 8 >> net.b{1}
 9
10 ans =
11
12        5.4365
13        5.4366
```

The weights are taking the values of the input vectors due to the following
steps:

```
1 % Extracting two features from two PPG signals
2 % The first signal is normotensive and the second
   signal is hypertensive
```

```
 3 Feature1 = [1.12 0.11];
 4 Feature2 = [0.7 0.33];
 5 % Competitive layer with two classes
 6 net1 = competlayer(2);
 7 % Combining features
 8 feature = [Feature1; Feature2];
 9 % training the competetive layer with the
     featureset
10 net = train(net1,feature);
11 % test the network with the same feature set used
     in the training
12 outputs = net(feature);
13 % converting the output into class label
14 classes = vec2ind(outputs);
```

Figure 9.1 shows an iterative process during the training phase of the neural network using the function *train*.

9.5 DISCRIMINANT ANALYSIS

The main idea of discriminant analysis is to project the data onto a line (or plane) that maximizes the distance between the classes. MATLAB has a powerful function that does this analysis and returns the classification as follows:

```
 1 class = classify(sample,training,group,'type') %
     classifies each row of the data in a sample into
     one of the groups in training. The sample and
     training must be matrices with the same number
     of columns. group is a grouping variable for
     training. Its unique values define groups; each
     element defines the group to which the
     corresponding row of training belongs. group can
     be a categorical variable, a numeric vector, a
     character array, a string array, or a cell array
     of character vectors. training and group must
     have the same number of rows. classify treats
     <undefined > values, NaNs, empty character
     vectors, empty strings, and <missing> string
     values in group as missing data values, and
     ignores the corresponding rows of training. The
     output class indicates the group to which each
     row of sample has been assigned, and is of the
     same type as group
```

- linear: fits a multivariate normal density to each group, with a pooled estimate of covariance; this is the default.

- diaglinear: similar to linear, but with a diagonal covariance matrix estimate (naive Bayes classifiers).

- quadratic: fits multivariate normal densities with covariance estimates stratified by group.

- diagquadratic: similar to quadratic, but with a diagonal covariance matrix estimate (naive Bayes classifiers).

- mahalanobis: uses Mahalanobis distances with stratified covariance estimates.

Assume that we have extracted four features from normotensive and hypertensive subjects, and that we need to classify the four features in two classes:

```
1  >> Feature1=[1.12 0.11];
2  >> Feature2=[0.7 0.33];
3  >> Feature3=[1.01 0.10];
4  >> Feature4=[0.6 0.3];
5  >> feature=[Feature1;Feature2;Feature3;Feature4];
6  >> Group=['Class 1';'Class 2';'Class 1';'Class 2'];
```

The *feature* matrix has been created, and categories were assigned to each feature vector. It is now time to classify feature vectors into two classes using the *classify* function:

```
1  >> Feature1=[1.12 0.11];
2  >> Feature2=[0.7 0.33];
3  >> Feature3=[1.01 0.10];
4  >> Feature4=[0.6 0.3];
5  >> feature=[Feature1;Feature2;Feature3;Feature4];
6  >> Group=['Class 1';'Class 2';'Class 1';'Class 2'];
7  >> [C,err,P,logp,coeff] = classify(feature,feature,
      Group,'linear')
```

The classification step was carried out and the result of the test features "feature" is saved in variable C. Note that we tested the classifier on the same training features, which will allow us to know whether the classifier is correct. As expected, the test features were classified exactly as the trained features with *err* = 0. The first feature and the third feature were classified as Class 1, while the second and fourth features were classified as Class 2, as follows:

```
 1  C =
 2
 3     4 7 char array
 4
 5        'Class 1'
 6        'Class 2'
 7        'Class 1'
 8        'Class 2'
 9
10
11  err =
12
13        0
14
15
16  P =
17
18        1.0000        0.0000
19        0.0000        1.0000
20        1.0000        0.0000
21        0.0000        1.0000
22
23
24  logp =
25
26        4.4301
27        4.4301
28        4.4301
29        4.4301
30
31
32  coeff =
33
34     2 2   struct array with fields:
35
36        type
37        name1
38        name2
39        const
40        linear
```

As mentioned above, the *classify* function allows different types of discriminant analysis. Let us now try a different typ—in this instance, "diagquadratic"—as follows:

```
1  >> Feature1=[1.12 0.11];
2  >> Feature2=[0.7 0.33];
3  >> Feature3=[1.01 0.10];
4  >> Feature4=[0.6 0.3];
5  >> feature=[Feature1;Feature2;Feature3;Feature4];
6  >> Group=['Class 1';'Class 2';'Class 1';'Class 2'];
7  >> [C,err,P,logp,coeff] = classify(feature,feature,
       Group,'diagquadratic')
```

The results were also correct with this function, with *err* = 0. The first feature and the third feature were classified as Class 1, while the other two features were classified as Class 2.

```
1  C =
2
3     4 7  char array
4
5        'Class 1'
6        'Class 2'
7        'Class 1'
8        'Class 2'
9
10
11 err =
12
13        0
14
15
16 P =
17
18        1.0000      0.0000
19        0.0000      1.0000
20        1.0000      0.0000
21        0.0000      1.0000
22
23
24 logp =
25
26        4.4746
27        3.4713
```

```
28      4.4746
29      3.4713
30
31
32  coeff =
33
34   2 2    struct array with fields:
35
36      type
37      name1
38      name2
39      const
40      linear
41      quadratic
```

This example shows that different types of discriminant analysis can lead to the same result. Note that if this were a real application, we would not be using the training feature matrix for testing. We used this concept to clarify the idea and make it easy to understand.

9.6 OTHER CLASSIFIERS

There are many classification methods that can be used. Four different classification methods have been discussed above. Other classifiers can be checked via the interactive MATLAB app *classificationLearner*.

The *classificationLearner* app can train and classify all features using multiple classifiers (e.g. decision trees, discriminant analysis, support vector machines, logistic regression, nearest neighbors, and ensemble classification, to mention a few). You can import your features and labels from workspace, select a feature set, specify the type of validation, train classifiers, and check all results. In fact, the app highlights the best classifier with highest degree of accuracy.

Let us use the example in the previous section to perform supervised machine learning using the *classificationLearner* app. First, the features and labels have to be combined into one variable called *data*, as follows:

```
1  >> Feature1=[1.12 0.11];
2  >> Feature2=[0.7 0.33];
3  >> Feature3=[1.01 0.10];
4  >> Feature4=[0.6 0.3];
5  >> feature=[Feature1;Feature2;Feature3;Feature4];
6  >> data=[feature [1; 2; 1; 2]];
```

Currently, the matrix *data* contains the features in the first two columns, while the labels are in the third column. It is now time to call up the learner app, as follows:

```
1 >> classificationLearner
```

Then, select the variable *data* from the list in the Workspace Variable. Notice that the app automatically detected that the last column contains the response (or labels). Also, select the type of validation, here a 4-fold cross-validation method was used, as shown in Figure 9.2.

After setting all the variables, press to start session. Then, select 'All' classifiers, afterwards clicking on "train". Once all the classifiers have been trained and validated, the best classifier result will be highlighted, as shown in Figure 9.3.

The "Data Browser" tap contains the accuracy for each classifier. You can check the result of each classifier. In total, 23 classifiers were tested and evaluated. If we are interested in knowing the behavior of the classifier, we can select the "Confusion Matrix" tap to see the number of false positives and false negatives, as shown in Figure 9.4.

FIGURE 9.2 Setting variables in the *classificationLearner* app.

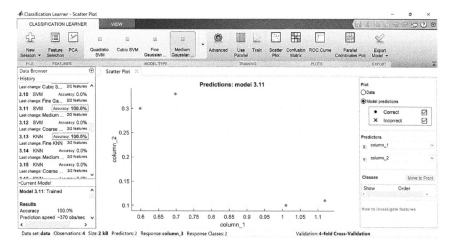

FIGURE 9.3   Feature scattering.

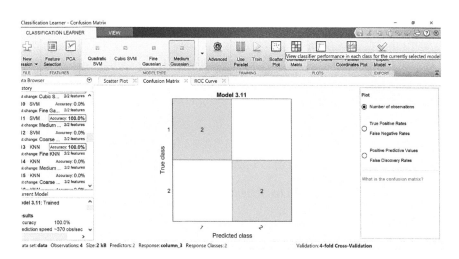

FIGURE 9.4   Confusion matrix of the best classifier shown in Figure 9.3.

Note that, given the features we have, the number of false positives and false negatives is zero. This is not usually the case in real applications. Moreover, the ROC curve is ideal, as shown in Figure 9.5, as the detection accuracy is 100%.

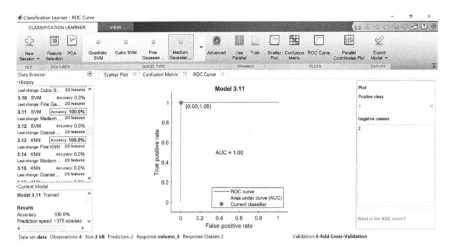

FIGURE 9.5 Receiver operating characteristic of the best classifier shown in Figure 9.3.

9.7 CLASSIFICATION EXAMPLE USING CLASSICAL MACHINE LEARNING METHODS

A mini example is presented to enable the reader to understand the basic workflow of classification easily. The dataset of this example is a matrix. The first column is the blood pressure labels and the other columns are the PPG features. The main steps of this example contain:

- Load dataset;

- Statistics analysis and data visualization for dataset;

- Using ReliefF method to do feature selection;

- Adopt the top N features to create a optimal feature set;

- Using 10-fold cross-validation;

- Building classifiers to train the trainset;

- Test the trained model;

- Undertake performance evaluation;

- Export the classification result.

```
1  %%%%%%%%%%%%%%%%%%%%%%%%%%%%%%%%%%%%%%%%%%%%%%%%%%%%%%%%
2  %Section 12.7.5
```

```
 3          % START_Classification .m
 4  %%%%%%%%%%%%%%%%%%%%%%%%%%%%%%%%%%%%%%%%%%%%
 5  %Notes:------------------BP label
 6  %Label value      1        2
 7  %Group          Norm       HT
 8
 9  %% set folder path
10  >> clc; clear; close all;
11
12  >> addpath(genpath(pwd));
13  >> path . data = ' . \DATA\ ' ; path . output =
       ' . \OUTPUT\ ' ;
14
15  %% Load classification dataset
16  >> dataset = load([path.data,'Norm_vs_HT.csv']);
17  >> target_Y = dataset(:,1);
18  >> feature_X = zscore(dataset(:,2:end));
19
20  %% Statistics and Data Visualization for dataset
21  >> [row,column]= size(dataset);
22  >> count.sample = row;
23  >> count.feature = column-1;
24  >> count.Norm = sum(target_Y==1);
25  >> >> count.HT = sum(target_Y==2);
26  >> figure(1); pie3([count.Norm,count.HT],{'Norm
       BP','HT BP'});legend(num2str(count.Norm),
       num2str(count.HT))
27
28  %% Feature Selection : ReliefF Method
29  >> [relieff_fList,relieff_weight] = relieff(feature
       _X,target_Y,10,'method','classification');
30  >> figure(2); bar(relieff_weight); xlabel('feature
       index');ylabel('feature weight');
31
32  %% Adopt the TopN valuable features to classify
33  >> TopN = floor(0.3*count.feature);
34  >> opt_feature_X = feature_X(:,relieff_
       fList(1:TopN));
35
36  %% 10-fold Cross-validation
37  >> Kfold = 10;
38  >> indices = crossvalind('Kfold',target_Y,Kfold);
       %10
39
40  >> sum_Acc = 0; sum_Sen = 0; sum_Spe = 0; sum_Pre =
       0; sum_Recall= 0;sum_F1 = 0;
```

```
41 >> for num_CV= 1:1:Kfold
42         test = (indices == num_CV); train = ~test;
43
44 >>      %% Divide the dataset into trainset (90%)
          and testset (10%)
45 >>      train_X = opt_feature_X(train,:);
46 >>      train_Y = target_Y(train,:);
47
48 >>      test_X = opt_feature_X(test,:);
49 >>      test_Y = target_Y(test,:);
50
51    %% Training Model
52 >>      classifier_type = 'KNN';
53 >> switch classifier_type
54       case 'LDA'      %Classifier 1: Linear
              Discriminant Analysis (LDA)
55        trainedModel = fitcdiscr(train_X,train_Y);
56       case 'KNN'       %Classifier 2: k-Nearest
              Neighbor
57        trainedModel = fitcknn(train_X,train_Y);
58       case 'SVM'        %Classifier 3: Support
              Vector Machine (SVM)
59        trainedModel = fitcsvm(train_X,train_Y);
60       case 'Tree'      %Classifier 4: Decision Tree
61        trainedModel = fitctree(train_X,train_Y);
62 >>    end
63
64    %% Testing Model
65 >>    [predicted_Y] = predict(trainedModel,test_X);
66
67    %% Performance Evaluation
68 >>    [stats] = classification_evaluation(test_Y,pr
          edicted_Y,2);
       % In Section 12.7.4
69
70 >>      sum_Acc = sum_Acc + stats.Acc;
71 >>      sum_Sen = sum_Sen + stats.Se;
72 >>      sum_Spe = sum_Spe + stats.Sp;
73 >>      sum_Pre = sum_Pre + stats.P;
74 >>      sum_Recall = sum_Recall + stats.P;
75 >>      sum_F1 = sum_F1 + stats.F1;
76 >> end
77
78 %% Averaged Performance Result
79 >> mean_Acc = sum_Acc/Kfold;
```

```
80  >> mean_Sen = sum_Sen/Kfold;
81  >> mean_Spe = sum_Spe/Kfold;
82  >> mean_Pre = sum_Pre/Kfold;
83  >> mean_Recall = sum_Recall/Kfold;
84  >> mean_F1 = sum_F1/Kfold;
85
86  >> head_name = {'Kfold','Norm','HT','Acc','Sen',
        'Spe','Pre','Recall','F1'};
87  >> T=table(Kfold, count.Norm, count.HT,...
        mean_Acc, mean_Sen, mean_Spe, mean_Pre,
        mean_Recall,mean_F1,...
88
89      'VariableNames',head_name,'RowNames',
        {classifier_type});
90  >> writetable(T,[path.output 'classification_
        result.csv'],'WriteRowNames',true);
```

9.8 CLASSIFICATION EXAMPLE USING DEEP LEARNING

Deep learning is a novel and powerful technology. With the increasing availability of computing resources, deep learning has become a popular solution in computer vision, text recognition, speech translation, physiological data mining, and so on. Here is a mini example of GoogLeNet transfer learning to show the working framework for deep learning. The dataset used in the example is our research dataset, which comprises PPG scalograms transformed by continuous wavelet transform (CWT). As discussed in Section 9.7.5, this dataset has two categories: Normaltensive (Norm) and Hypertension (HT).

```
 1  %%%%%%%%%%%%%%%%%%%%%%%%%%%%%%%%%%%%%%%%%%%%%%%
 2  %Section 12.7.6
 3  % START_Classification_DeepLearning .m
 4  %%%%%%%%%%%%%%%%%%%%%%%%%%%%%%%%%%%%%%%%%%%%%%%
 5  %Notes:--------BP label
 6  %Label value      1        2
 7  %categories      Norm      HT
 8
 9  %% set folder path
10  >> clc; clear; close all;
11
12  >> addpath(genpath(pwd));
13  >> path.data = ' . \DATA\ '; path . output =
        ' . \OUTPUT\ ';
```

```
14
15 %% Load image classification dataset into
     imageDatastore
16 >> parentDir = path.data;
17 >> dataDir = 'data';
18 >> allImages = imageDatastore(fullfile(parentDir,
     dataDir),'IncludeSubfolders',true,'LabelSource',
     'foldernames');
19
20 %% Statistics and Data Visualization for dataset
21 >> count.Norm = sum(str2num(char(allImages.Labels))
     == 1);
22 >> count.HT = sum(str2num(char(allImages.Labels))
     == 2);
23 >> figure(1); pie3([count.Norm,count.HT],{'Norm
     BP','HT BP'});legend(num2str(count.Norm),
     num2str(count.HT))
24
25 %% Divide the dataset into trainset (60%) ,
     validation (20%) and testset (20%)
26 >> [imgsTrain,imgsValidation,imgsTest] = splitEach
     Label(allImages,0.6,0.2,0.2,'randomized');
27 >> count.train = numel(imgsTrain.Files);
28 >> count.validation = numel(imgsValidation.Files);
29 >> count.test = numel(imgsTest.Files);
30
31 %% Build Transfer Learning Model using GoogLeNet
32 %Notice : GoogLeNet requires RGB images of size
     224-by-224-by-3.
33 >> net = googlenet;
34
35 %Extract the layer graph from the network and plot
     the layer graph .
36 >> lgraph = layerGraph(net);
37
38 %% Modify GoogLeNet Network Parameters
39 >> lgraph = removeLayers(lgraph,{'pool5-drop_
     7x7_s1','loss3-classifier','prob','output'});
40
41 >> numClasses = numel(categories(imgsTrain.
     Labels));
42 >> newLayers = [
43        dropoutLayer(0.6,'Name','newDropout')
```

```
44          fullyConnectedLayer(numClasses,'Name',
            'fc',' WeightLearnRateFactor',5,
            'BiasLearnRateFactor',5)
45          softmaxLayer('Name','softmax')
46          classificationLayer('Name','classoutput')];
47 >> lgraph = addLayers(lgraph,newLayers);
48
49 >> lgraph = connectLayers(lgraph,'pool5-7x7_s1',
               'newDropout');
50
51 %% Training Options Setting
52 >> options = trainingOptions('sgdm',...
53       'MiniBatchSize',64,...
54       'MaxEpochs',3,…
55       'InitialLearnRate',1e-4,...
56       'ValidationData',imgsValidation,...
57       'ValidationFrequency',10,...
58       'ValidationPatience',Inf,...
59       'Verbose',1,...
60       'ExecutionEnvironment','cpu',...
61       'Plots','training-progress');
62
63 %% Training Model
64 >> trainedGN = trainNetwork(imgsTrain,lgraph,
     options);
65
66 %% Testing Model
67 >> [YPred,probs] = classify(trainedGN,imgsTest);
68
69 %% Performance Evaluation
70 >> [stats] = classification_evaluation(str2num(char
     (imgsTest.Labels)),str2num(char(YPred)),2);
71
72 %% Export Classification Results
73 >> head_name = {'Norm','HT','train','validation',
     'test','Acc','Sen','Spe','Pre','Recall','F1'};
74 >> T=table(count.Norm, count.HT,count.train, count.
     validation,count.test, ...
75       stats.Acc, stats.Se, stats.Sp, stats.P,
          stats.Recall,stats.F1, ...
76       'VariableNames',head_name,'RowNames',{'GoogLe
       Net'});
77 >> writetable(T,[path.output 'classification_
     result.csv'],'WriteRowNames',true);
```

9.9 EFFECTIVENESS EVALUATION

A very easy mistake to make when attempting to evaluate the effectiveness of your "product" or "design" is testing it on the same feature set that you used for its development. The goal of evaluation is to get an idea of how well it may perform in practice and the real world (also known as its "generalizability"). Something that works extremely well in development but has poor generalization essentially has no usel as a product.

To test its generalizability without deploying it in the real world, you can withhold a subset of your feature set from training to be used later for evaluation. Since the features have not been seen or used during training, it simulates a real-world application. Overall, a feature set can be separated into the following categories:

- **Training**: used exclusively for training/developing your algorithms.

- **Validation**: used during the development phase to validate hyperparameter choices, and so on, or it can be used for evaluating objective functions during optimization. For example, when training a machine learning algorithm over n-iterations, performance on the validation set can be used to identify early-stopping points

- **Test**: a part of the total feature set that has not been used for training/development, and is solely used to get an idea of how well it will perform on unknown and new features.

If the initial feature set has a test set withheld, with development and training using the remaining set of features, the evaluation method is called 'hold-out' testing, which is the simplest form of validation.

9.9.1 K-Fold Cross Validation

K-fold cross validation (K-CV) expands the simple hold-out by dividing the feature set into K non-overlapping equally sized subsets. This way, the algorithm can be trained and evaluated K-times. For k from 0 to $k - 1$, the test set is the k-th subset while the algorithm is trained on the remaining subsets. This way, you can obtain the average performance over K times, which provides a good indication of the generalizability of the algorithm itself in the real world.

MATLAB provides a simple function for separating your features into k-subsets that you can then use to employ cross-fold validation, or for separating into simple train/validation/test. This function can be found in Section 9.6.

9.9.2 Class Imbalance

It is possible that a feature set is class-imbalanced. When a classifier is trained on a class-imbalanced feature set, the classifier will tend to learn more about the class with the greater number of samples, and therefore lean towards classifying things as the over-sampled class. A simple way to solve this is to resample your features, randomly discarding samples of the class with the greater number of samples so that the final feature set you use for training/development is no longer class-imbalanced.

9.9.3 Confusion Matrix

A test feature set, such as the training or validation feature set, will consist of a set of features for each sample, and their true class labels. Evaluation of performance will then be performed by comparing the predictions made by your algorithm when given the test set samples and those of their true class labels. A confusion matrix is a simple graphical method for showing you the results of the evaluation, as shown in Figure 9.6.

A confusion matrix shows the frequency of samples classified correctly (output class = target class), and those that were not classified correctly. The confusion matrix discussed above is for a binary classifier. Let Class 0

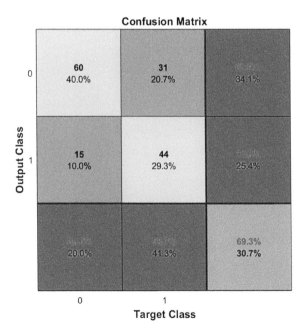

FIGURE 9.6 A MATLAB confusion matrix, plotting the results of evaluating an algorithm.

be a positive sample, and Class 1 be a negative sample; the confusion matrix will tell you the following information:

- True Positives (TP): Target Class Positive, Output Class Positive

- True Negatives (TN): Target Class Negative, Output Class Negative

- False Positives (FP): Target Class Negative, Output Class Positive

- False Negatives (FN): Target Class Positive, Output Class Negative

The *classificationLearner* app provides a confusion matrix as shown in Section 9.6. MATLAB also provides functions that will plot the confusion matrix (confusionmat or plotconfusion), or summarize classifier performance (called classperf), for which it requires only the true class labels and the predicted class labels from your test set. See Example 2 for generating a confusion matrix and Example 1-1 for classifier performance.

```
1  %% Example 2
2  % Create features .
3  t = zeros(150,2);
4  y = zeros(150,2);
5  t(1:75,1) = 1;
6  t(76:150,2) = 1;
7  y(1:44,1) = 1;
8  y(44:75,2) =1;
9  y(76:90,1) =1;
10 y(90:150,2) =1;
11 % t is the ONE-HOT true labels matrix, for binary
       classes 150 samples
12 % y is the ONE-HOT classifier predictions, for
       binary classes 150 samples
13
14 % Using confusionmat. Confusionmat needs label
       VECTORS, and not One-hot
15 % matrices. First convert the one-hot labels to
       label vector. Since we have
16 % a binary classifier , let us use the first column
       as the label vector
17 conmat_1 = confusionmat(t(:,1),y(:,1));
18 disp(conmat_1)
19
20 % To plot a fancy looking confusion matrix like the
       one in the Tutorial,
```

```
21 % use plotconfusion.
22 %   plotconfusion also uses label vectors, but
     requires each column to be one sample
23 %   , so we transpose the column vector into a row
     vector
24 figure;
25 plotconfusion(t(:,1)',y(:,1)');
```

9.9.4 Sensitivity versus Specificity

The sensitivity (or true positive rate) of a classifier is calculated by the following equation:

$$TPR = \frac{TP}{TP + FN}$$

which is the probability that, given that all samples are positive, the classifier will detect them as positive.

The specificity (or true negative rate) of a classifier is calculated by:

$$TNR = \frac{TN}{TN + FP}$$

which is the probability of a classifier predicting a negative condition when all samples are actually negative.

Sensitivity and specificity are commonly used for medical evaluations because these functions are *independent* of the prevalence of the condition being classified in population, and solely reflect the abilities of the test/algorithm itself.

Unlike sensitivity and specificity, positive/negative predictive values (and their derived metrics) are *dependent* on the prevalence of the condition in a population. Here is a MATLAB function that can help with evaluating an application's performance:

```
1 % This function calculates the sensitivity and
     specificity
2 % and other evaluation measures
3 function [stats] = classification_evaluation
     (TARGET,PREDICTED,positive_label_value)
4
5 % Input :
6 % -------
```

```
 7 % TARGET = Column matrix with TARGET class labels
                      of the training
 8 %                 examples
 9 % PREDICTED = Column matrix with predicted class
                      labels by the
10 %                  classification model
11
12 % Output :
13 % --------
14 % stats is a structure array
15 % stats . confusionMat
16 %            Predicted Classes
17 %                 p '      n '
18 %           ___| _____ | _____ |
19 %    Actual p |         |        |
20 %    Classes n |         |        |
21 %
22 % Four important parameters :
23 % --------------------------
24 % TP: true positive
25 % TN: true negative
26 % FP: false positive
27 % FN: false negative
28 % --------------------
29 % Calculating different evaluation measures
30 % -----------------------------------------
31 % stats . accuracy = (TP + TN) /(TP + FP + FN + TN)
32 % stats . precision = TP / (TP + FP)
33 % stats . sensitivity = TP / (TP + FN)
34 % stats . specificity = TN / (FP + TN)
35 % stats . recall = sensitivity
36 % stats . Fscore = 2*TP /(2*TP + FP + FN)
37
38 index = (TARGET() == positive_label_value);  %index
       of data with positive class
39 num_positive = length(TARGET(index)); num_negtive =
       length(TARGET(~index));
40 N = num_positive + num_negtive;
41 TP = sum(TARGET(index) == PREDICTED(index));
42 TN = sum(TARGET(index) == PREDICTED(index));
43 FP = num_negtive-TN;
44 FN = num_positive-TP;
45
46 TPR = TP/(TP+FN); FNR = FN/(TP+FN); FPR = FP/
       (FP+TN); TNR = TN/(TN+FP);
```

```
47
48 accuracy = (TP+TN)/(TP + FN + FP + TN);
49 sensitivity = TPR;
50 specificity = TNR;
51 precision = TP/(TP+FP);
52 recall = sensitivity;
53 F1 = 2*((precision*recall)/(precision + recall));
54
55 stats.positive_label = positive_label_value;
56 stats.TP = TP;
57 stats.FP = FP;
58 stats.FN = FN;
59 stats.TN = TN;
60 stats.Acc = accuracy;
61 stats.Se = sensitivity;
62 stats.Sp = specificity;
63 stats.P = precision;
64 stats.Recall = recall;
65 stats.F1 = F1;
66
67 stats.All = [positive_label_value accuracy TP FP FN
        TN TPR FNR FPR TNR precision sensitivity
        specificity F1];
68
69 end
```

9.10 SUMMARY

The Bayes classifier for normally distributed features is quadratic, and linear in the case of equal covariance matrices. Also, the minimum Mahalanobis distance classifier is optimum for normally distributed features, equal covariance matrices, and equal priors. Both Euclidean and Mahalanobis distance classifiers are linear classifiers. There are many classifiers available in MATLAB. A validation phase is required to provide a robust classifier. It is recommended that the learner app is used to select the classifier that provides the highest degree of accuracy, and, finally, a check based on the confusion matrix and the ROC curve is required.

CHAPTER 10

Application of PPG to Global Health

This chapter discusses potential opportunities for the use of PPG signals in global health. Recently, the World Health Organization established a project called the "WHO Patient Safety Pulse Oximetry," to improve the safety of anaesthesia care in operating rooms in low and middle-income countries. Pulse oximetry at the moment offers only heart rate and oxygen saturation levels, however it is expected that the PPG signals will be extracted at some point to provide a diagnostics value (to detect hypertensive subjects, etc) to improve the safety of operating rooms worldwide.

10.1 LEARNING OBJECTIVES

The learning objectives of this chapter are to:

- Develop an understanding of the need for PPG signals in global health

- Learn about the six-step framework for global health approach

- Gain insight into three noncommunicable disease case studies

In 2012, the World Health Organization (WHO) estimated that non-communicable diseases (NCDs) accounted for 68% of 56 million total deaths worldwide. Of these NCD deaths, approximately 80% (28 million) occurred in low- and middle-income countries (LMICs).[143] Cardiovascular diseases (17.5 million deaths, or 46% of all NCD deaths); cancers (8.2 million, or 22% of all NCD deaths); and respiratory diseases, including asthma and chronic obstructive pulmonary disease (4.0 million) were the leading causes of these deaths.[144]

The physiological processes of these diseases and the consequential deaths to which they led produce changes in the human body. These changes are accompanied by biomedical signals, and these signals can reflect the nature and degree of the change. Deaths caused by noncommunicable diseases present physiological parameters that differ when compared to the parameters of healthy individuals. The prediction, prevention, and treatment of NCDs (in their early stages) is possible through the analysis of these signals. Recent advances in biomedical signal collection and analysis, mobile diagnostic and screening devices, and machine learning applications presents a feasible and practical solution to tackling NCDs in a variety of settings, including LMICs.

Medical diagnoses and scientific hypotheses are supported by the detection, storage, transmission, and analysis of the valuable information provided by biomedical signal analysis. This process of analysis presents information via a rich visualization and interpretation method that allows clinicians and researchers to utilize the information for practical medical and research application. Across various scientific disciplines and health care industries, a multitude of methods are being applied that aim to produce efficient, convenient, and robust tools. Mobile health (mHealth) is of particular interest to current industry leaders, as it is anticipated that mobile devices can be used to overcome barriers of access to health care and to improve diagnostic accuracy.[145,146]

Current research aims to assist in the prediction and diagnosis of diseases through the meaningful extraction of data that can then be used to create PPG-based point-of-care (POC) technologies. Knowledge yielded from this approach can support the development of therapies that can be used earlier in the course of a disease, consequently improving treatment, reducing morbidity, and improving overall quality of life. Two important aspects must also be considered: scalability and affordability are essential for practical application and deployment of any tools developed, otherwise utilization of the developed technologies will be impractical and unattainable.

The proposed PPG-based framework discusses six steps that are essential to guiding the scientific development process, beginning with the essential first step of developing a scientific hypothesis, and ending with practical application. The PPG framework is comprises the following steps: *simplicity, mining, connection, reliability, affordability, and scalability* (SMCRAS). Applying the full framework during the research and design process for PPG-based devices facilitates the development process for researchers. Through this process, researchers are provided the opportunity to examine their developments and investigate how they can be extended for larger-scale and global use.

Based on a current literature scan, the proposed framework for PPG-based POC devices is the first of its kind. Recently, in 2014, researchers began to explore the relationship between biomedical engineering and how application of this science can improve global health.[145] In 2015, researchers outlined an affordability framework for POC devices,[146] which is one element of the proposed framework. Two PPG-based case studies (with technical examples) are presented to illustrate the real-world application of SMCRAS and how its application relates to global health improvement.

10.3 OVERVIEW

Signals are a significant component in our everyday life and play a major role in our daily communication. A variety of signals are interpreted and processed by humans on a day-to-day basis, including verbal, social, mental, and physical signals. Each of these signal types holds valuable information that gives us a deeper understanding of situations, contexts, and of each other as humans. Biomedical signals, like any other signal, bear important information that, if analyzed and interpreted appropriately, gives us a deeper understanding of the human body and its inner workings. Body parts and their related biosignals provide information about the state of that specific part, as well as the whole body. Specifically, like any communication system dealing with signals, the biosignal involve a sender, receiver, and a medium through which the signal is sent. Algorithms can be used to understand these signals, which, in turn, can help us screen and detect abnormality in the body. Mobile phones are becoming more extensively used to collect biomedical signals, often to aid clinicians and health care practitioners with the prediction, diagnosis, and treatment of disease.

While mobile devices are a promising tool to support this data collection, there are several challenges regarding data collection. Inconsistent user training (which impacts signal quality collection), inconsistent measurement (which impacts signal quality), limited computational resources (which impacts processing and transmission), finite power (which impacts usability), time constraints for clinical staff (which impacts overall collection), varying user interface designs (which impacts accessibility and usability), and uncontrolled environments (which impacts noise interference for signal collection), all factor into the robustness and quality of mobile device use. The SMCRAS framework factors these limitations into the development of PPG-based algorithms/devices, in addition to providing a guideline for how the the devices can be used for the classification and prediction of disease.

The six steps that address the main objectives are to: (i) aim for simplicity, (ii) mine through noise based on information detection theory, (iii) reveal hidden connections, (iv) assess robustness, (v) plan for scalability, and (vi) strive for affordability. Each objective is discussed in detail below.

10.4 SIMPLICITY

Simplicity was described by Einstein when he commented: "Any intelligent fool can make things bigger and more complex. It takes a touch of genius and a lot of courage to move in the opposite direction." In terms of mobile computation, simplicity is particularly effective when high detection accuracy is achieved with less storage and power. [119] Electrocardiogram (ECG) signal QRS complex detection is accomplished by both;[55,110] however, the simpler algorithm requires fewer execution steps and this is a less complex way to achieve the same result.

PPG signals are collected via POC devices (either wirelessly or wired) and the information is then sent for further analysis to a central monitoring station using a global system for mobile communications (GSM) or an Internet connection.[147,148] Depending on the device, part of the analyses is executed on the POC device locally before transmission. In these cases, the transmission often consumes more power than the signal analysis itself,[149,150] making it a less favorable option. Algorithms used for real-time analysis must therefore retain simplicity (to conserve energy, require less time, and so on,) given that such simplicity does not significantly decrease accuracy. Theoretically, when applied properly, a simpler algorithm will result in the faster processing of large-scaled biomedical signals. Moreover, it will require less power in POC devices, and this is particularly important with devices that are battery-driven.[119] At times, there can be a trade-off between the simplicity and accuracy of an algorithm, since higher accuracy can sometimes be achieved by a more complex algorithm. The aim, therefore, is to develop a comparably more simple algorithm that is reliable and accurate for tackling large data.[150] Application of the "simple" approach can be investigated on PPG-based POC devices for NCD and global health care screening and diagnostics.

10.5 MINING

The term "mining" is used to describe informative pattern extraction from datasets, which often varies from discipline to discipline (e.g. data mining is used as information/knowledge extraction, information/knowledge discovery, information/knowledge harvesting, or data analysis/processing). The more preferred term used in the field of biomedical engineering is "Data mining", which will be used throughout this text. For the purposes of this section, the "mining" step is defined as a combination of filtering and feature extraction phases.

Mining noisy biomedical signals can be challenging, and often requires precision and patience: at times, it can be tempting to give up before obtaining an informative waveform. Filters are often used to clean the noise from the signal, which consequently compromises the integrity of the waveform. Thus, it is imperative to develop filtering methods that preserve the main waveforms of the processed signal, thereby protecting the valuable data and information in the original waveform. This improved filtering method yields a more accurate detection method that consequently improves the prediction of disease screening and diagnosis.

Providing real-time feedback on signal quality to health care workers is also imperative, as it enables the instantaneous recollection of signals if this is required due to the initial collection of a poor quality signal. Combined with an improved filtering method, real-time feedback will improve clinical practice by providing more informative signals, which, in turn, provides more accurate information for clinicians to screen and diagnose diseases.

10.6 CONNECTION

Mining for hidden features and analyzing them in conjunction with the relatively more obvious features will reveal relationships that would otherwise be overlooked. Once extracted, correlations and causalities for the medical abnormality (i.e. disease of interest) can be calculated using these various features (both obvious and hidden) from the biomedical signal.

The analysis of statistical and deterministic properties of biomedical signals needs to begin with the assessment of the following features: kurtosis, skewness, energy, entropy, line length/curve length, minima/maxima, activity (1st Hjorth parameter), mobility (2nd Hjorth parameter), complexity (3rd Hjorth parameter), RMS amplitude, zero crossings, and relative power. Biomarkers are then created by correlating the extracted features with the NCD disease of interest. After rigorous testing of these biomarkers, machine-learning algorithms can be created for the automatic detection of these biomarkers. Alternatively, practitioners can be trained to identify the biomarkers.

10.7 RELIABILITY

"Simplicity is a prerequisite for reliability"—Edsger Dijkstra. In other words, one must exist in order for the other to exist. After completing the first three steps of the SMCRAS framework (one of which is achieving simplicity), the developed solution needs to be tested for reliability and accuracy. Following through with this step is necessary for the algorithm/device to hold up against current international standards. Furthermore, this step provides evidence that the PPG-based POC solution performs as well, or even better than, current comparable diagnostic tools.

Reliability testing is determined by using four specific and individual components: true positives, true negatives, false positives, and false negatives. Several statistical measures can then be calculated using these four values, which, in turn, assess the reliability of the algorithm (e.g. its sensitivity, specificity, and positive predictivity).

Before deploying the PPG-based POC device, quality control must also be implemented to detect and prevent errors. This step is in addition to the typical quality control and risk management reviews completed during the manufacturing process. Health care professionals must also go through individual on-site quality control testing before mass use to confirm the device is appropriate for the intended purpose. Systematic management of quality is also necessary and can be carried out through creating organizational procedures, ongoing user training and evaluation, and periodic evaluation of management principles.

10.8 AFFORDABILITY

Health care quality varies greatly between high-income and low-income countries, and these differences are typically due to poor infrastructure, lack of qualified practitioners, challenges to accessibility, and difficulty with managing costs. In recent years, due to an increase in the implementation of mobile broadband networks in LMICs (89%), mobile devices are increasingly becoming the device of choice for the collection of biomedical signals. This approach also addresses some of the health care challenges mentioned above; the use of mobile devices addresses accessibility (through the conversion of mobile phones into PPG-based POC devices), cost management (sensors are low-cost items that can be hooked into these mobile devices), and the lack of trained practitioners (devices can be programmed for automated detection and therefore a wider range of health care workers can be trained on the device).

PPG-based POC devices offer an affordable and feasible solution for serving larger populations and for improving health care outcomes. Affluent populations already use PPG-based POC devices (e.g. heart rate monitoring via wearable watches and smartphones). While most of these devices and apps focus on heart rate monitoring and step-counting, it is feasible to expect that this same technology can also be used to tackle more serious conditions and diseases. The use of mobile devices is on the rise; internationally, there is a paradigm shift towards developing and implementing solutions that incorporate global health principles (affordability, accessibility, aiding LMICs, and so on) for preventative, diagnostic, and treatment purposes. The SMCRAS framework complements and integrates well into this shift, seeking to address unmet needs in both privileged and underprivileged/remote areas and countries with vulnerable populations.[146]

10.9 SCALABILITY

Intuitively, it can be said that simplicity can ensure the likelihood of scalability for PPG-based POC devices and related algorithms (i.e. simple algorithms require less complex devices, and less complex devices are easier to reproduce and to train a wider range of practitioners in their use). While drawing this conclusion is partially accurate, it only considers part of the full definition of scalability in the framework. At times, simple devices are designed to be used on a specific patient type and in a specific environment. Scalability shifts the focus from these two restrictions and elevates device application to a broader community and patient focus. By creating mobile solutions for widescale application in a variety of settings and on different patient subsets, we can achieve scalability globally.

User-friendly design is also an essential component that impacts scalability. Clear, directive instructions, coupled with a pleasing user interface and a system that is easy to use, will increase use and uptake by a variety of practitioners. Additionally, the algorithm/solution needs to be designed so that it can be used with different applications and devices. Providing algorithms that work across different sampling frequencies without needing an adjustment for a particular sampling frequency, or for a specific parameter and condition increases the likelihood of scalability.

The measure of success for scalability includes being able to use the device across a variety of settings, with a variety of users, without major complications or restrictions (e.g. remote sites, in-clinic, on different patient subsets). Mass implementation of technology through the SMCRAS framework is only achievable when each of the subsequent steps is followed and the process is finalized by the scalability step. When scalability is achieved, knowledge sharing is at its best and many are able to reap the benefits of the mobile solution, impacting health outcomes globally.

10.10 NONCOMMUNICABLE DISEASE CASE STUDIES

10.10.1 Case I: Detection of Heat Stress in a Changing Climate

Anthropogenic greenhouse gas emissions are one of the main causes of accelerated global warming, according to the Intergovernmental Panel on Climatic Change. Symptoms of this warming include increased frequency and intensity of heatwaves, and higher mean ambient temperatures.[151] Heat stress, which is the net heat load to which an individual is exposed, is impacted by both of these factors. Body heat dissipation, which is essential for cooling down the body, is limited in warmer environments. Workers in labor-intensive industries who endure high levels of physical exertion, and consequently high levels of body heat generation, are especially susceptible to heat stress[152] due to the impact of global warming.[153] Heat stroke, permanent harm, and sometimes even death are risks associated with sustained heat stress.[154] During heatwave periods, it is approximately 4–7 times more likely that workers will experience the impact of heat on their health.[155] Additionally, injury claims in workers are positively related to ambient temperature.[156]

Labor-intensive workers are at high-risk for heat stress, thus mitigation could include screening for heat tolerance[157] and physiological monitoring.[152] Studies have identified two physiological heat stress indices that are measured during physical activity: body core temperature (BCT) and heart rate (HR). The standard measurement site for BCT assessment is the rectum,[158] but due to its invasive nature alternative sites have been explored. The forehead, temporal, oral, aural, and axilla temperature serve as alternatives, although the indices collected from these sites following activity are not as accurate as the standard.[159] These issues and limitations are problematic in many occupational settings.

Heart rate measurement, due to its less complex nature in terms of measurement, provides an alternative means for heat stress measurement. Hot climates will induce significantly higher HR whether a worker is seated and at rest [160,161] or performing physical activity.[162] In hotter climates, the narrowed scope for body heat dissipation causes an elevated HR, which, in turn, increases BCT and skin temperature. Heart rate variability (HRV) is the analysis of changes in the heart rate and is quantified as the reduced root mean square of the differences of successive differences; thus HRV indices are proposed as an objective indicator of heat stress.[163] HRV calculation typically requires ECG signals recordings of considerable length. Thus, creating a simple, non-invasive method for in-the-field assessment of heat stress is needed, thereby allowing the monitoring of workers during their working hours. Achieving this can prevent heat stress symptoms and heat-related illnesses and deaths.

10.10.1.1 Simplicity

Prediction of heat stress for susceptible individuals requires invasive diagnostic testing that assesses maximal aerobic power and/or heat tolerance in a controlled setting.[156,164] Specialized equipment is used in a controlled climate, and is accompanied by the induction of high levels of physiological strain. Moreover, BCT and HRV physiological monitoring typically requires specialized equipment that is invasive and requires complex analysis. PPG signal collection offers a non-invasive alternative that is simple to measure. The signal is collected from a fingertip during scheduled breaks at work, and recent improvements in wearable sensors enable the continuous measurement of PPG. Analysis of the PPG collected provides insight into physiological strain, heat stress, and autonomic arousal. Techniques for heat stress analysis have been deter-mined in the literature,[69] whereas HRV standards of heat stress have yet to be developed.

10.10.1.2 Mining

Preservation of the saliency of systolic and diastolic waves is possible through the use of the bandpass filter and the dicrotic notch. Recent research has recommended the use of a zero-phase second-order Butterworth filter with a bandpass of 0.5–8 Hz to remove the baseline wander and high frequencies.[53] The detection of systolic peaks in PPG signals measured after exercise presents challenges caused by motion arti-facts, sweat, and nonstationary effects.[53] While some studies have exam-ined filters and algorithms to analyze the PPG wave contour, the methods lack accuracy and reproducibility.[165] Consequently, researchers are begin-ning to apply the second derivative for the easy quantification and empha-sis of the subtle changes in the PPG contour;[66] the second derivative can also be used to improve the mining step, while increasing accuracy.

10.10.1.3 Connection

Recent attempts to connect PPG features of heat stress with global warm-ing[69] have been made. Furthermore, new features to determine the opti-mal PPG feature for heat stress detection are being explored, such as the energy of the aa area. A total of 14 time-domain features—seven PPG sig-nal features and seven APG signal features[69]—were tested to see how these concepts connect the concepts of PPG signal feature extraction, heat stress detection, and global warming.

10.10.1.4 Reliability

A combination of entropy and an HRV index was used in the algorithm in Elgendi et al. (2015),[69] achieving a sensitivity of 95% and positive predic-tivity of 90.48% on 40 healthy, heat-acclimatized emergency responders (30 males and 10 females) with a median age of 34 years. Participants were normotensive (had a mean systolic blood pressure of 129.3 mmHg, range 110–165 mmHg) and had no known cardiovascular, neurological, or respiratory disease. The range of systolic blood pressure exceeded the usual normotensive range, and the results were considered reliable and more robust when compared to the existing method.

PPG signals were collected in a noisy setting and analyzed on a laptop computer offline. Given the challenging environment and the promise of the results, these methods have demonstrated promise for PPG-based POC implementation.

10.10.1.5 Affordability

PPG devices used in developing countries are quite costly, totaling approximately US$300 per unit.[166] Conversion of these devices into a mobile phone that is used as a PPG-based POC device is an alternative mobile solution. When comparing the cost of this device to the cost of a mobile phone in some developing countries, the total is approximately US$15, with the additional cost of the PPG sensor being approximately only US$3 more. The less expensive option of the mobile phone is promising in terms of affordability, especially when compared against the current devices in use.

10.10.1.6 Scalability

The use of PPG signals for the monitoring of anesthesia in surgery has been the standard of care in developed countries for more than 20 years, and the WHO is leading a project that aims to make PPG monitoring available in every operating room across the globe.[167] Use of the PPG probe is intuitive, requiring minimal knowledge. The probe clip is simply placed on the patient's finger for the collection of the PPG signal. Software on the device then shows the PPG signal in real time along with the automatic diagnosis.

10.10.2 Case II: Prediction of Adverse Outcomes Related to Preeclampsia using SpO2

This case study shows how the blood oxygen saturation level (SpO2) can indicate the pregnancy disorder preeclampsia (PE), which is characterized by high blood pressure and proteinuria. Worldwide, it effects approximately 3–8% of all pregnancies and is responsible for approximately 18.5% of maternal deaths each year.[168] The burden of PE is disproportionately felt in LMICs, where it is estimated that PE causes approximately 99% of annual maternal and perinatal deaths.[169]

10.10.2.1 Simplicity

Hypertension and proteinuria must both be present in order to diagnose PE;[170] thus, two separate instruments are required to measure these disease characteristics for diagnosis. At 20 weeks in the gestation period, a blood pressure cuff is used to check whether the subject's blood pressure is 140 mm Hg (systolic), or ≥90 mm Hg (diastolic) in women with previously normal blood pressure. A urine test is also needed to check whether proteinuria is ≥ 0.3 g of protein in a 24-hour urine collection.[170] In developing countries, these two tests are commonly unavailable; thus, there is a need to improve (or replace) current diagnostic procedures with simple methodology.

By calculating the ratio of the AC components and the ratio of the DC components of the two light sources, as shown in Figure 2.1, SpO2 can be obtained from each pulse of a PPG signal. SpO2 saturation is related to hypertension,[171] and, thus, collecting the PPG signal, which carries SpO2 information, by using the pulse oximeter is a simpler way to diagnose/detect PE and its related complications.

10.10.2.2 Mining
Please refer to Case 1; the same methodology applies.

10.10.2.3 Connection
The risk prediction index is calculated using SpO2, since it is linked to hypertension.[171] The PPG probe emits lights from the light-emitting diode (LED), which can be detected either on the same side (reflectance mode) or on the opposite side (transmittance mode) of the tissue by a photodetector. Output from the photodetector is converted into a voltage and further processed, producing a PPG signal.[172] The PPG signal can be divided into a pulsatile (alternating current/ac) component and a relatively constant (direct current/dc) component. SpO2 is then calculated using the ac and dc ratio components of the red and infrared PPG signals together with a calibration curve.[172,173,174] SpO2 values are then calculated every 10 seconds ac, and the dc PPG amplitudes are determined using the empirically calibrated equation.[173,174] The reliability of SpO2 has been tested in a proof-of-concept study through the miniPIERS prediction model,[175] which included a cohort of 726 women (118 of whom had experienced adverse pregnancy outcomes) in South Africa and Pakistan. Women with either suspected or confirmed PE were admitted into hospital. Preliminary results demonstrated that adding SpO2 derived from PPG signals improved prediction accuracy from 81% to 84%.[171] PPG signals were collected via a mobile phone that had a real-time analysis capability, which is an example of a real-world practical implementation for PPG-based POC devices.

10.10.2.4 Affordability
Please refer to Case 1; the same methodology applies.

10.10.2.5 Scalability

Please refer to Case 1; the same methodology applies.

10.10.3 Case III: Hypertension Risk Stratification

Hypertension or high blood pressure is linked to 1,100 deaths per day in the United States alone; therefore, it is critical that regularly people at risk are regularly monitored. The challenge is that current blood pressure monitoring methods have their limitations.

10.10.3.1 Simplicity

The gold standard for the assessment of hypertension is the intra-arterial blood pressure measurement, which is highly accurate but invasive, requiring the doctor to insert a needle into an artery. The other non-invasive alternative the sphygmomanometer (manual or electronic). This measurement method has been widely recognized and popularized over the last century of development and it has played a major role in the control of cardiovascular diseases. However, a sphygmomanometer can easily be affected by the conditions of it operation and use, such as the operation of the cuff, the sitting posture of the patient, and exercise. In addition, it also has a definite white coat phenomenon (blood pressure is higher when it is taken in a medical setting than when it is taken at home) for some patients. Therefore, new cuff-less hypertension screening and blood pressure detection technologies are needed. PPG signals can offer an alternative.

10.10.3.2 Mining

There are two mining approaches to investigating this topic: a feature-based approach[176–178] or a non-feature-based approach.[179]

10.10.3.3 Connection

For the feature-based approach, many features can be examined in order to test whether the features are correlated with hypertension. In a study[179] published in 2019, a total of 125 features were extracted. These features covered time span (23 features), PPG amplitude (14 features), VPG and APG features (10 features), the waveform area (4 features), the power area (15 features), ratio (43 features), and slope (16 features). The results were promising, as the accuracy were high: detection of normotension versus prehypertension achieved 72.97%, normotension and prehypertension versus hypertension achieved 81.82%, and normotension versus, achieved 92.31%.

For the non-feature-based approach, PPG signals are processed without extracting features. In a study[178] published in 2018, the results were promising: the accuracy achieved for detecting normotension versus prehypertension was 80.52%; for normotension and prehypertension versus hypertension, 92.55%; and for normotension versus hypertension, 82.95%.

10.10.3.4 Affordability
Please refer to Case 1 and Case 2; the same methodology applies.

10.10.3.5 Scalability
Please refer to Case 1 and Case 2; the same methodology applies.

10.11 USER PERFORMANCE
For global health approaches, collecting PPG signals from different countries is essential. This presents many challenges, such as those experienced by users who are collecting PPG signals in remote areas. Users in these circumstances are not usually expert in using pulse oximetry; neither are

MONTHLY STATISTICS - JUNE 2015

1651 Recordings

131 Users

Signal Quality Index (SQI)
Below are the recordings broken down into three quality categories

Application Quality
Below are the recordings broken down into three quality categories

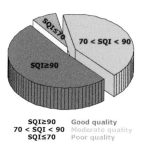

SQI≥90	Good quality
70 < SQI < 90	Moderate quality
SQI≤70	Poor quality

FIGURE 10.1 A sample of the monthly report shared by the UBC Central team with the CLIP Pakistan team on the quality of SpO2 recordings. *Source*: Adapted from the pulse oximetry. PRE-EMPT Annual Report (p. 23).[180]].

they familiar with the amount of noise associated with the data collection phase. Therefore, encouraging users and providing feedback to them on a regular basis would play an essential role for the collection of high-quality data, consequently improving diagnostic decisions.

An example of the feedback I was generating on a monthly basis for users collecting PPG signals from Pakistan is shown in Figure 10.1. As can be seen, having a variety of different sites is ideal for global health research; unfortunately, the various locations increases the likelihood of noise due to environmental and other uncontrolled issues.

As mentioned above, my focus during this project has been to assess the quality of PPG signals, provide feedback to sites, and investigate different

FIGURE 10.2   A report sent to the CLIP Pakistan team of the monthly statistics of Lady Health Worker (LHW) performance in a measurement of SpO2. *Source*: Adapted from the pulse oximetry. PRE-EMPT Annual Report (p. 24).[180]].

quality indices that will enable us to create an app to deploy an algorithm inside the miniPIERS app that helps the user collect better quality signals.

As part of the evaluation process, we looked at a number of different aspects to help sites collect better signals via feedback training, and to determine signal quality. We also assessed challenges related to body posture, finger/hand/body movement, hand sweat and dirt, nail polish, partial closure of probe, environmental noise, and software functionality.

I ranked the user performance in a monthly report shared with the sites (see Figures 10.1 and 10.2 present an example of a monthly report). The reports were well-received by the sites and the level of performance has increased. When data are compared between two consecutive months, such as May and June, it is clear that there was an increase (from 42% to 51%) in the high-quality signals, as shown in Figure 10.2.

Finally, developing robust feature detection algorithms and SQIs for your PPG-based application will improve not only the feedback to users, but also the data collection process in general.

10.12 SUMMARY

The analysis and processing of PPG signals has immense potential to play a major role in early disease detection. Current barriers to accessing health care in LMICs can be addressed with advances in technology, offering solutions that are based on proposed PPG-based signals and the SMCRAS framework. Identifying inexpensive and non-invasive alternatives to diagnostic devices that rely on signal processing will help to reduce morbidity, mortality, and disability rates, particularly in developing countries. SMCRAS proposes a road map to biomedical signal analysis and implementation that can be used across many disciplines. The SMCRAS framework comprises six sequential steps: simplicity, mining, connection, reliability, affordability, and scalability. Two case studies have been presented that demonstrate the real-world application of the SMCRAS framework. When applied correctly, the framework further facilitates the development of PPG-based POC technologies, which will have a significant impact on mortality and morbidity rates, especially for those living in LMICs. Providing feedback to users is essential for global health research, especially for users with minimal knowledge.

CHAPTER 11

Available PPG Databases

This chapter highlights different PPG datasets that are publicly available and can be downloaded to start applying all methods discussed in the previous chapters. To date, these datasets are the most known PPG datasets. It is recommended that you download any of these datasets and start wrangling with the PPG signals to benefit the most from this book.

11.1 FINGERTIP PPG FROM HYPERTENSIVE SUBJECTS

The incidence of cardiovascular disease (CVD) has risen around the world in recent years, overtaking the mortality rate of cancer, thus making CVD the number one killer of humans. Many studies have been conducted using non-invasive early diagnosis and screening techniques for CVDs, such as hypertension and coronary artery sclerosis, in order to discover more convenient and effective methods for the early identification of CVDs. Of these methods, the PPG has become widely recognized as a low-cost non-invasive detection technology for CVDs. The cardiovascular parameters detected using PPG technology include heart rate, blood oxygen saturation, blood pressure, assessment of arterial stiffness, and pulse wave velocity, among others. The PPG signal includes information on the hemodynamic process, hemorheology, and tissue status of the peripheral microcirculation system in the human body. Thus, it can be seen that the PPG signal is an aggregated expression of many physiological processes in the cardiovascular circulation system. A physiological information database with a high level of precision and a high sampling rate is urgently needed in PPG technology research in order to extract a greater number of cardiovascular parameters for the early screening and diagnosis of CVDs.

We provide, here, a database containing physiological information and PPG waveform data collected over one year that can be used to research arterial blood vessel aging, arterial blood pressure detection, and screening of hypertensive and diabetic patients based on PPG signals.

This PPG and BP (PPG-BP) database integrates the de-identified, comprehensive clinical data of patients admitted to the Guilin People's Hospital in Guilin, China. The openness of the data allows clinical studies to explore and improve the understanding of relationships between cardiovascular health and PPG signals, the final goal being to create a simple, effective, non-invasive detection technology that is easy to use and wearable. This dataset has been collected from 219 subjects, aged 21–86 years, with a median age of 58 years. Males accounted for 48%. The dataset covers several diseases, including hypertension, diabetes, cerebral infarction, and insufficient brain blood supply.

In summary, this unique non-invasive detection dataset for cardiovascular disease can be used in a wide range of in-depth research.

- Further details and illustrative example signals can be found in the reference paper:

  Liang Y, Chen Z, Liu G, Elgendi M. A new, short-recorded photoplethysmogram dataset for blood pressure monitoring in China. *Scientific Data*. 2018 Feb 27; 5: 180020.
- To download the PPG-BP database, please go to this website: https://figshare.com/articles/PPG-BP_Database_zip/5459299

11.2 FINGERTIP PPG FROM AN INTENSIVE CARE UNIT

The MIMIC II Waveform Database contains thousands of recordings of multiple biosignals and time series of vital signs collected from bedside patient monitors in adult and neonatal intensive care units (ICUs). It is a companion to the MIMIC II Clinical Database, which contains detailed clinical information for many of the patients represented in the Waveform Database. The MIMIC II Waveform Database Matched Subset contains 4,897 waveform records and 5,266 numerical records from the MIMIC II Waveform Database, which have been matched and time-aligned with 2,809 MIMIC II Clinical Database records.

Note that all recorded signals vary depending on choices made by the ICU staff. Therefore, there is no consistent measurement in terms of the number of collected biosignals. Some records contain one or more ECG signals, and often include continuous arterial blood pressure signals, fingertip PPG signals, and respiration, with additional biosignals (up to 8 biosignals may be collected simultaneously). The recording length also varies; most are of a few days in duration, but some are shorter and others are several weeks long.

It is worth noting, as it is a large dataset, that the MIMIC II Waveform Database has not been annotated. Moreover, the MIMIC database contains biosignals that were collected simultaneously but not synchronized. Many papers have been published with results using the MIMIC database assuming that all the signals were collected at the same time and synchronized. A recent paper[176] discussed a feature that relies on the synchronicity between PPG and ECG using the MIMIC database, and also discussed features that do not rely on synchronicity extracted from PPG waveforms. However, the researchers[181] who created the database have stated that there are errors in the data matching and alignment, indicating that some (perhaps all) recordings are not synchronized.

- Further details and illustrative example signals can be found in the reference paper:

  Saeed M, Villarroel M, Reisner AT, Clifford G, Lehman LW, Moody G, Heldt T, Kyaw TH, Moody B, Mark RG. Multiparameter Intelligent Monitoring in Intensive Care II (MIMIC-II): a public-access intensive care unit database. *Critical Care Medicine*. 2011 May; 39(5): 952.
- To download the MIMIC-II database, please go to this website: https://physionet.org/physiobank/database/mimic2wdb/

11.3 WRIST PPG DURING EXERCISE

This database contains wrist PPGs recorded during walking, running, and bike riding. Accelerometers and gyroscope signals were used remove some of the motion interference from the PPG traces. In addition, the ECG signals were collected to be used as a reference (or gold standard) for heart rate estimation during exercise.

Measurements were taken using an ECG unit placed on the chest, together with a PPG and Inertial Measurement Unit placed on the left wrist while participants used an indoor treadmill and exercise bike. A single ECG channel (using two electrodes) was recorded using an Actiwave (CamNtech, Cambridge, UK) recorder and pre-gelled self-adhesive Silver-Silver Chloride (Ag/AgCl) electrodes. These were placed on the upper chest, with one electrode on either side of the heart. R peaks in this ECG trace were identified by hand and these timings are included in the database to allow a gold standard reference heart rate comparison.

A Shimmer sensing system (Dublin, Ireland) was used. The system contains a PPG probe, gyroscope, accelerometer, and magnetometer. The PPG sensor was glued to the main Shimmer unit in order to give a rigid connection and to allow the movement sensors inside the main Shimmer unit to make an accurate recording of the movement of the PPG sensor. The combined unit was then placed on the left wrist in the approximate position of a standard watch. Participants were asked to perform one or more different types of exercise. Four options were available:

- walking on a treadmill at a normal pace for up to 10 minutes;

- a light jog/run on a treadmill, at a pace set by the participant, for up to 10 minutes;

- pedaling on an exercise bike set at a low resistance for up to 10 minutes;

- pedaling on an exercise bike set at a higher resistance for up to 10 minutes.

The goal of the study was to collect PPG signals in a very noisy environment. As such, each participant was free to set the pace of the treadmill and the pedal rate on the bike so that they were comfortable; they were also free either to change these settings, or to stop the exercise at any time. Most participants spent between 4 and 6 minutes on each activity. In all cases, the subject was starting from rest. All signals were sampled at 256 Hz. The group comprised 8 participants: 3 male and 5 female subjects aged between 22 and 32 years (giving the group a mean age of 26.5 years).

- Further details and illustrative example signals can be found in the reference paper:

  Jarchi D, Casson AJ. Description of a database containing wrist PPG signals recorded during physical exercise with both accelerometer and gyroscope measures of motion. *Data*. 2016 Dec 24; 2(1): 1.
- To download the wrist PPG during examination of the exercise database, please go to this website: https://physionet.org/physiobank/database/wrist/

11.4 FINGERTIP PPG AND RESPIRATION

Originally, the data were collected from critically ill patients during hospital care at the Beth Israel Deaconess Medical Centre (Boston, MA, USA). Fortunately, this database is annotated by two annotators who manually annotated individual breaths in each recording using the impedance respiratory signal. The 53 recordings within the dataset, each of 8 minutes duration, each contain:

- physiological signals, such as the PPG, impedance respiratory signal, and electrocardiogram (ECG). These are sampled at 125 Hz.

- physiological parameters, such as the heart rate (HR), respiratory rate (RR), and blood oxygen saturation level (SpO2). These are sampled at 1 Hz.

- fixed parameters, such as age and gender.

- manual annotations of breaths taken.

The goal of this study was to evaluate the performance of different algorithms for estimating the respiratory rate from PPG signals.

- Further details and illustrative example signals can be found in the reference paper:
  Pimentel MA, Johnson AE, Charlton PH, Birrenkott D, Watkinson PJ, Tarassenko L, Clifton DA. Toward a robust estimation of respiratory rate from pulse oximeters. *IEEE Transactions on Biomedical Engineering*. 2017 Aug 1; 64(8): 1914–23.
- To download the PPG and respiration database, please go to this website: https://physionet.org/physiobank/database/bidmc/

11.4.1 The University of Queensland Vital Signs Dataset

The University of Queensland Vital Signs dataset contains a wide range of patient monitoring data and vital signs that were recorded during 32 surgical cases where patients underwent anaesthesia at the Royal Adelaide Hospital.

- Further details and illustrative example signals can be found in the reference paper:

Liu D, Gorges M, Jenkins, SA. The University of Queensland vital signs dataset: development of an accessible repository of anesthesia patient monitoring data for research. *Anesthesia & Analgesia*. 2012; 114(3): 584–9.

- To download the University of Queensland Vital Signs dataset, please go to this website: https://outbox.eait.uq.edu.au/uqdliu3/uqvitalsignsdataset/index.html

11.4.2 BioSec.Lab PPG Dataset

The BioSec.Lab PPG dataset was created for research inyo PPG-based biometrics recognition at the University of Toronto. This dataset includes signals recorded in four different conditions to evaluate the permanence, robustness, and uniqueness of the PPG signal as a biometric identity. The dataset contains signals recorded in the following settings: a relaxed state, after exercise, short time lapse, and long time lapse.

Relaxed state: a PPG signal of 3 minutes' duration recorded from 86 subjects. After exercise: a PPG signal of 3 minutes' duration recorded from 40 subjects after heavy exercise. Short time lapse: a PPG signal of 3 minutes' duration recorded in two parts at least 30 minutes apart on same day from 55 subjects. Long time lapse: a PPG signal of 3 minutes' duration recorded in two parts at least 2 weeks apart from 37 subjects. In addition, a fingertip video recording was made using a mobile camera: a video recording of 1 minute's duration of the fingertip from 36 subjects.

- Further details and illustrative example signals can be found in the reference paper:
    Yadav U, Abbas, SN, Hatzinakos D. Evaluation of PPG biometrics for authentication in different states. The 11th IAPR International Conference on Biometrics (ICB), Goldcoast, Australia, Feb 2018. (Preprint)
- To download the BioSec.Lab PPG Dataset, please go to this website: https://www.comm.utoronto.ca/~biometrics/PPG_Dataset/data_desc.html

11.4.3 Vortal Dataset

The Vortal dataset is a dataset of ECG and PPG signals acquired from healthy volunteers. This dataset is intended to facilitate evaluation of the performance of algorithms under ideal operating conditions.

- Further details and illustrative example signals can be found in the reference paper:

  Charlton PH, Bonnici TB et al. An assessment of algorithms to estimate respiratory rate from the electrocardiogram and photoplethysmogram. *Physiological Measurement.* 2016; 37(4): 610–26.

- To download the BioSec.Lab PPG Dataset, please go to this website: http://peterhcharlton.github.io/RRest/vortal_dataset.html

11.5 SUMMARY

The four PPG databases used in our discussion of feature extraction, feature selection, and feature detection are publicly available. These are:

- https://figshare.com/articles/PPG-BP_Database_zip/5459299

- https://physionet.org/physiobank/database/mimic2wdb/

- https://physionet.org/physiobank/database/wrist/

- https://physionet.org/physiobank/database/bidmc/

It is recommended that you download these databases and explore the morphological differences between the PPG waveform and its derivatives. In addition, try to analyze and visualize the most informative signals using time and frequency domain processing techniques. These databases are valuable for testing and evaluating your algorithm before deployment, as they were collected from subjects with a variety of health conditions and in various environments.

References

[1] Shelley K, Shelley S (2001) Pulse oximeter waveform: photoelectric plethysmography. In *Clinical Monitoring*, C Lake, R Hines, and C Blitt, Eds: WB Saunders Company , pp. 420–428.

[2] Allen J (2007) Photoplethysmography and its application in clinical physiological measurement. *Physiological Measurement* 28: R1.

[3] Millasseau S, Kelly R, Ritter J, Chowienczyk P (2002) Determination of age-related increases in large artery stiffness by digital pulse contour analysis. *Clinical Science* 103: 371–377.

[4] Padilla J, Berjano E, Saiz J, Facila L, Diaz P, et al. (2006) Assessment of relationships between blood pressure, pulse wave velocity and digital volume pulse. In: *IEEE Computers in Cardiology*. IEEE, pp. 893–896.

[5] Alty SR, Angarita-Jaimes N, Millasseau SC, Chowienczyk PJ (2007) Predicting arterial stiffness from the digital volume pulse waveform. *IEEE Transactions on Biomedical Engineering* 54: 2268–2275.

[6] Mishima N, Kubota S, Nagata S (1999) Psychophysiological correlates of relaxation induced by standard autogenic training. *Journal of Psychotherapy and Psychosomatics* 68: 207–213.

[7] Yashima K, Sasaki T, Kageyama Y, Odagaki M, Hosaka H (2005) Application of wavelet analysis to the plethysmogram for the evaluation of mental stress. In: *The 27th Annual International Conference of the Engineering in Medicine and Biology Society*. pp. 2781–2784.

[8] Kageyama Y, Odagaki M, Hosaka H (2007) Wavelet analysis for quantification of mental stress stage by finger-tip photo-plethysmography. In: *The 29th Annual International Conference of the IEEE Engineering in Medicine and Biology Society*. pp. 1846–1849.

[9] Abe M, Yoshizawa M, Sugita N, Tanaka A, Chiba S, et al. (2008) A method for evaluating effects of visually-induced motion sickness using ICA for photoplethysmography. In: *The 30th Annual International Conference of the IEEE Engineering in Medicine and Biology Society*. IEEE, pp. 4591–4594.

[10] Cox P, Madsen C, Ryan KL, Convertino VA, Jovanov E (2008) Investigation of photoplethysmogram morphology for the detection of hypovolemic states. In: *The 30th Annual International Conference of the IEEE Engineering in Medicine and Biology Society*. IEEE, pp. 5486–5489.

[11] Gil E, Mendez M, Vergara JM, Cerutti S, Bianchi AM, et al. (2009) Discrimination of sleep-apnea-related decreases in the amplitude fluctuations of ppg signal in children by hrv analysis. *IEEE Transactions on Biomedical Engineering* 56: 1005–1014.

[12] Takazawa K, Tanaka N, Fujita M, Matsuoka O, Saiki T, et al. (1998) Assessment of vasoactive agents and vascular aging by the second derivative of photoplethysmogram waveform. *Hypertension* 32: 365–370.

[13] Imanaga I, Hara H, Koyanagi S, Tanaka K (1998) Correlation between wave components of the second derivative of plethysmogram and arterial distensibility. *Japanese Heart Journal* 39: 775–784.

[14] Baek HJ, Kim JS, Kim YS, Lee HB, Park KS (2007) Second derivative of photoplethysmography for estimating vascular aging. In: *The 6th International Special Topic Conference on Information Technology Applications in Biomedicine.* IEEE, pp. 70–72.

[15] Kimura Y, Takamatsu K, Fujii A, Suzuki M, Chikada N, et al. (2007) Kampo therapy for premenstrual syndrome: efficacy of kamishoyosan quantified using the second derivative of the fingertip photoplethysmogram. *Journal of Obstetrics and Gynaecology Research* 33: 325–332.

[16] Katsuki K, Yamamoto T, Yuuzu T, Tanaka H, Okano R, et al. (1993) A new index of acceleration plethysmogram and its clinical physiological evaluation. *Nihon seirigaku zasshi Journal of the Physiological Society of Japan* 56: 215–222.

[17] Takada H, Washino K, Harrell JS, Iwata H (1995) Acceleration plethysmography to evaluate aging effect in cardiovascular system—Using new criteria of four wave patterns. *Medical Progress through Technology* 21: 205–210.

[18] Bortolotto A, Jacques B, Takeshi K, Kenji T, Michel S (2000) Assessment of vascular aging and atherosclerosis in hypertensive subjects: second derivative of photoplethysmogram versus pulse wave velocity. *American Journal of Hypertension* 13: 165–171.

[19] Ushiroyama T, Kajimoto Y, Sakuma K, Ueki M (2005) Assessment of chilly sensation in japanese women with laser doppler fluxmetry and acceleration plethysmogram with respect to peripheral circulation. *Bulletin of the Osaka Medical College* 51: 76–84.

[20] Nousou N, Urase S, Maniwa Y, Fujimura K, Fukui Y (2006, December). Classification of acceleration plethysmogram using self-organizing map. In 2006 *International Symposium on Intelligent Signal Processing and Communications.* IEEE, pp. 681–684.

[21] Taniguchi K, Nishikawa A, Nakagoe H, Sugino T, Sekimoto M, et al. (2007) Evaluating the surgeon's stress when using surgical assistant robots. In: *RO-MAN 2007-The 16th IEEE International Symposium on Robot and Human Interactive Communication.* IEEE, pp. 888–893.

[22] Fujimoto Y, Yamaguchi T (2008) Evaluation of mental stress by analyzing accelerated plethysmogram applied chaos theory and examination of welfare space installed user's vital sign. *IFAC Proceedings* 41: 8232–8235.

[23] Millasseau SC, Kelly RP, Ritter JM, Chowienczyk PJ (2003) The vascular impact of aging andvasoactive drugs: comparison of twodigital volume pulse measurements. *American Journal of Hypertension* 16: 467–472.

[24] Rivas-Vilchis JF, Hernández-Sánchez F, González-Camarena R, Suárez-Rodríguez LD, Escorcia-Gaona R, et al. (2007) Assessment of the vascular effects of pc6 (neiguan) using the second derivative of the finger photoplethysmogram in healthy and hypertensive subjects. *The American Journal of Chinese Medicine* 35: 427–436.

[25] Tokutaka H, Maniwa Y, Gonda E, Yamamoto M, Kakihara T, et al. (2009) Construction of a general physical condition judgment system using acceleration plethysmogram pulse-wave analysis. In: *International Workshop on Self-Organizing Maps.* Springer, pp. 307–315.

[26] Alnaeb ME, Alobaid N, Seifalian AM, Mikhailidis DP, Hamilton G (2007) Optical techniques in the assessment of peripheral arterial disease. *Current Vascular Pharmacology* 5: 53–59.

[27] Hertzman AB (1938) The blood supply of various skin areas as estimated by the photoelectric plethysmograph. *American Journal of Physiology–Legacy Content* 124: 328–340.

[28] Reisner A, Shaltis PA, McCombie D, Asada HH (2008) Utility of the photoplethysmogram in circulatory monitoring. *The Journal of the American Society of Anesthesiologists* 108: 950–958.

[29] Nilsson L, Johansson A, Kalman S (2005) Respiration can be monitored by photoplethysmography with high sensitivity and specificity regardless of anaesthesia and ventilatory mode. *Acta Anaesthesiologica Scandinavica* 49: 1157–1162.

[30] Rhee S, Yang BH, Asada HH (2001) Artifact-resistant power-efficient design of finger-ring plethysmographic sensors. *IEEE Transactions on Biomedical Engineering* 48: 795–805.

[31] Jung J, Lee J (2008) Zigbee device access control and reliable data transmission in zigbee based health monitoring system. In: *Advanced Communication Technology, 2008. ICACT 2008. 10th International Conference on.* IEEE, volume 1, pp. 795–797.

[32] Maeda Y, Sekine M, Tamura T (2011) Relationship between measurement site and motion artifacts in wearable reflected photoplethysmography. *Journal of Medical Systems* 35: 969–976.

[33] Poh MZ, Swenson NC, Picard RW (2010) Motion-tolerant magnetic earring sensor and wireless earpiece for wearable photoplethysmography. *IEEE Transactions on Information Technology in Biomedicine* 14: 786–794.

[34] Budidha K, Kyriacou P (2014) The human ear canal: investigation of its suitability for monitoring photoplethysmographs and arterial oxygen saturation. *Physiological Measurement* 35: 111.

[35] Wang L, Lo BP, Yang GZ (2007) Multichannel reflective ppg earpiece sensor with passive motion cancellation. *IEEE Transactions on Biomedical Circuits and Systems* 1: 235–241.

[36] Kyriacou PA (2013) Direct pulse oximetry within the esophagus, on the surface of abdominal viscera, and on free flaps. *Anesthesia & Analgesia* 117: 824–833.

[37] Mendelson Y, Pujary C (2003) Measurement site and photodetector size considerations in optimizing power consumption of a wearable reflectance pulse oximeter. In: Engineering in Medicine and Biology Society, 2003. *Proceedings of the 25th Annual International Conference of the IEEE.* IEEE, volume 4, pp. 3016–3019.

[38] Lee E, Shin J, Hong J, Cha E, Lee T (2010) Glass-type wireless ppg measuring system. In: *2010 Annual International Conference of the IEEE Engineering in Medicine and Biology.* IEEE, pp. 1433–1436.

[39] Tur E, Tur M, Maibach HI, Guy RH (1983) Basal perfusion of the cutaneous microcirculation: measurements as a function of anatomic position. *Journal of Investigative Dermatology* 81: 442–446.

[40] Tang Q, et al. (2020) PPGSynth: An innovative toolbox for synthesizing regular and irregular photoplethysmography waveforms. *Frontiers in Medicine* 7. doi:10.3389/fmed.2020.597774.

[41] Tang Q, et al. (2020) Synthetic photoplethysmogram generation using two Gaussian functions. *Scientific Reports* 10.1: 1–10.

[42] Verkruysse W, Svaasand LO, Nelson JS (2008) Remote plethysmographic imaging using ambient light. *Optics Express* 16: 21434–21445.

[43] Poh MZ, McDuff DJ, Picard RW (2010) Non-contact, automated cardiac pulse measurements using video imaging and blind source separation. *Optics Express* 18: 10762–10774.

[44] Poh MZ, McDuff DJ, Picard RW (2011) Advancements in noncontact, multiparameter physiological measurements using a webcam. *IEEE Transactions on Biomedical Engineering* 58: 7–11.

[45] Jonathan E, Leahy M (2010) Investigating a smartphone imaging unit for photoplethysmography. *Physiological Measurement* 31: N79.

[46] Elgendi M, Fletcher R, Liang Y, Howard N, Lovell NH, et al. (2019) The use of photoplethysmography for assessing hypertension. *NPJ Digital Medicine* 2: 1–11.

[47] Allen J, Murray A (2000) Variability of photoplethysmography peripheral pulse measurements at the ears, thumbs and toes. *IEE Proceedings-Science, Measurement and Technology* 147: 403–407.

[48] Korhonen I, Yli-Hankala A (2009) Photoplethysmography and nociception. *Acta Anaesthesiologica Scandinavica* 53: 975–985.

[49] Elgendi M, Jonkman M, De Boer F (2009) Measurement of *a-a* intervals at rest in the second derivative plethysmogram. In: *The IEEE Conference in Bioelectronics and Bioinformatics*, pp. 75–79.

[50] Elgendi M, Jonkman M, De Boer F (2010) Heart rate variability measurement using the second derivative photoplethysmogram. In: *The 3rd International Conference on Bio-inspired Systems and Signal Processing* (BIOSIGNALS2010): pp. 82–87.

[51] Elgendi M, Norton I, Brearley M, Abbott D, Schuurmans D (2014) Detection of *a* and *b* waves in the acceleration photoplethysmogram. *Biomedical Engineering Online* 13: 139.

[52] Elgendi M (2014) Detection of *c*, *d*, and *e* waves in the acceleration photoplethysmogram. *Computer Methods and Programs in Biomedicine* 117: 125–136.

[53] Elgendi M, Norton I, Brearley M, Abbott D, Schuurmans D (2013) Systolic peak detection in acceleration photoplethysmograms measured from emergency responders in tropical conditions. *PLoS ONE* 8: e76585.

[54] Elgendi M, Jonkman M, De Boer F (2010) Frequency bands effects on QRS detection. In: *The 3rd International Conference on Bio-inspired Systems and Signal Processing* (BIOSIGNALS2010): pp. 428–431.

[55] Elgendi M (2013) Fast QRS detection with an optimized knowledge-based method: evaluation on 11 standard ECG databases. *PLoS ONE* 8: e73557.

[56] Elgendi M, Eskofier B, Abbott D (2015) Fast T wave detection calibrated by clinical knowledge with annotation of P and T waves. *Sensors* 15: 17693.

[57] Elgendi M, Kumar S, Guo L, Rutledge J, Coe JY, et al. (2015) Detection of heart sounds in children with and without pulmonary arterial hypertension-daubechies wavelets approach. *PLoS ONE* 10: e0143146.

[58] Elgendi M (2016) Eventogram: a visual representation of main events in biomedical signals. *Bioengineering* 3: 22.

[59] Braunwald E, Zipes D, Libby P, Bonow R (2004) *Braunwald's Heart Disease: A Textbook of Cardiovascular Medicine*, 7th edition. Philadelphia: Saunders.

[60] Liang Y, Elgendi M, Chen Z, Ward R (2018) An optimal filter for short photoplethysmogram signals. *Scientific Data* 5: 180076.

[61] Yoon G, Lee JY, Jeon KJ, Park KK, Yeo HS, et al. (2002) Multiple diagnosis based on photoplethysmography: Hematocrit, SpO2, pulse, and respiration. *Proceedings of the SPIE* 4916: 185–188.

[62] Lu S, Zhao H, Ju K, Shin K, Lee M, et al. (2008) Can photoplethysmography variability serve as an alternative approach to obtain heart rate variability information? *Journal of Clinical Monitoring and Computing* 22: 23–29.

[63] Elgendi M (2012) On the analysis of fingertip photoplethysmogram signals. *Current Cardiology Reviews* 8: 14–25.

[64] Shelley K (2007) Photoplethysmography: beyond the calculation of arterial oxygen saturation and heart rate. *Anesthesia & Analgesia* 105: S31–S36.

[65] Petterson MT, Begnoche VL, Graybeal JM (2007) The effect of motion on pulse oximetry and its clinical significance. *Anesthesia & Analgesia* 105: S78–S84.

[66] Seki H (1977) Classification of wave contour by first and second derivative of plethysmogram (in Japanese). *Pulse Wave* 7: 42–50.

[67] Takazawa K, Fujita M, Kiyoshi Y, Sakai T, Kobayashi T, et al. (1993) Clinical usefulness of the second derivative of a plethysmogram (acceleration plethysmogram). *Cardiology* 23: 207–217.

[68] Elgendi M, Fletcher RR, Norton I, Brearley M, Abbott D, et al. (2015) Frequency analysis of photoplethysmogram and its derivatives. *Computer Methods and Programs in Biomedicine* 122: 503–512.

[69] Elgendi M, Norton I, Brearley M, Fletcher RR, Abbott D, et al. (2015) Towards investigating global warming impact on human health using derivatives of photoplethysmogram signals. *International Journal of Environmental Research and Public Health* 12: 12776.

[70] Hartmut Gehring M, Me HM, Schmucker P (2002) The effects of motion artifact and low perfusion on the performance of a new generation of pulse oximeters in volunteers undergoing hypoxemia. *Respiratory Care* 47: 48–60.

[71] Cannesson M, Delannoy B, Morand A, Rosamel P, Attof Y, et al. (2008) Does the pleth variability index indicate the respiratory-induced variation in the plethysmogram and arterial pressure waveforms? *Anesthesia & Analgesia* 106: 1189–1194.

[72] Colquhoun D, Forkin K, Durieux M, Thiele R (2012) Ability of the masimo pulse co-oximeter to detect changes in hemoglobin. *Journal of Clinical Monitoring and Computing* 26: 69–73.

[73] Shan C (2013) Motion robust vital signal monitoring. US Patent App. 14/031,451.

[74] Krishnan R, Natarajan B, Warren S (2010) Two-stage approach for detection and reduction of motion artifacts in photoplethysmographic data. *Biomedical Engineering, IEEE Transactions on* 57: 1867–1876.

[75] Selvaraj N, Mendelson Y, Shelley K, Silverman D, Chon K (2011) Statistical approach for the detection of motion/noise artifacts in photoplethysmogram. In: *2011 Annual International Conference of the IEEE Engineering in Medicine and Biology Society.* pp. 4972–4975. doi:10.1109/IEMBS.2011.6091232.

[76] Clifford G, Behar J, Li Q, Rezek I (2012) Signal quality indices and data fusion for determining clinical acceptability of electrocardiograms. *Physiological Measurement* 33: 1419.

[77] Cohen J (1960) A coefficient of agreement for nominal scales. *Educational and Psychological Measurement* 20: 37–46.

[78] Tong S, Li Z, Zhu Y, Thakor N (2007) Describing the nonstationarity level of neurological signals based on quantifications of time-frequency representation. *IEEE Transactions on Biomedical Engineering* 54: 1780–1785.

[79] Coifman R, Wickerhauser M (1992) Entropy-based algorithms for best basis selection. *IEEE Transactions on Information Theory* 38: 713–718.

[80] Chen Ch (1988) *Signal Processing Handbook*, volume 51. CRC Press.

[81] Vaseghi SV (2008) *Advanced Digital Signal Processing and Noise Reduction.* John Wiley & Sons.

[82] Li BN, Dong MC, Vai MI (2010) On an automatic delineator for arterial blood pressure waveforms. *Biomedical Signal Processing and Control* 5: 76–81.

[83] Billauer E (2012) peakdet: Peak detection using MATLAB, http://billauer.co.il/peakdet.html.

[84] Elgendi M (2016) Optimal signal quality index for photoplethysmogram signals. *Bioengineering* 3: 21.

[85] Hosanee M, Chan G, Welykholowa K, Cooper R, Kyriacou PA, et al. (2020) Cuffless single-site photoplethysmography for blood pressure monitoring. *Journal of Clinical Medicine* 9: 723.

[86] Elgendi M, Liang Y, Ward R (2018) Toward generating more diagnostic features from photoplethysmogram waveforms. *Diseases* 6: 20.

[87] Asada HH, Shaltis P, Reisner A, Sokwoo R, Hutchinson RC (2003) Mobile monitoring with wearable photoplethysmographic biosensors. *IEEE Engineering in Medicine and Biology Magazine* 22: 28–40.

[88] Chua CP, Heneghan C (2006) Continuous blood pressure monitoring using ecg and finger photoplethysmogram. In: *The 28th Annual International Conference of the IEEE Engineering in Medicine and Biology Society*, New York: pp. 5117–5120.

[89] Murray W, Foster P (1996) The peripheral pulse wave: information overlooked. *Journal of Clinical Monitoring and Computing* 12: 365–377.

[90] Dorlas JC, Nijboer JA (1985) Photo-electric plethysmography as a monitoring device in anaesthesia: application and interpretation. *British Journal of Anaesthesia* 57: 524–530.

[91] Chua ECP, Redmond SJ, McDarby G, Heneghan C (2010) Towards using photo-plethysmogram amplitude to measure blood pressure during sleep. *Annals of Biomedical Engineering* 38: 945–954.

[92] Awad AA, Haddadin AS, Tantawy H, Badr TM, Stout RG, et al. (2007) The relationship between the photoplethysmographic waveform and systemic vascular resistance. *Journal of Clinical Monitoring and Computing* 21: 365–372.

[93] Seitsonen E, Korhonen I, Van Gils M, Huiku M, Lötjönen J, et al. (2005) Eeg spectral entropy, heart rate, photoplethysmography and motor responses to skin incision during sevoflurane anaesthesia. *Acta Anaesthesiologica Scandinavica* 49: 284–292.

[94] Wang L, Pickwell-MacPherson E, Liang Y, Zhang YT (2009) *Noninvasive cardiac output estimation using a novel photoplethysmogram index*. In: *Annual International Conference of the IEEE Engineering in Medicine and Biology Society*, Minneapolis, pp. 1746–1749.

[95] Linder SP, Wendelken SM, Wei E, McGrath SP (2006) Using the morphology of photoplethysmogram peaks to detect changes in posture. *Journal of Clinical Monitoring and Computing* 20: 151–158.

[96] Fu TH, Liu SH, Tang KT (2008) Heart rate extraction from photoplethysmogram waveform using wavelet multi-resolution analysis. *Journal of Medical and Biological Engineering* 28: 229–232.

[97] Gil E, Orini M, Bailón R, Vergara J, Mainardi L, et al. (2010) Photoplethysmography pulse rate variability as a surrogate measurement of heart rate variability during non-stationary conditions. *Physiological Measurement* 31: 1271.

[98] Poon C, Teng X, Wong Y, Zhang C, Zhang Y (2004) Changes in the photoplethysmogram waveform after exercise. In: *Computer Architectures for Machine Perception, 2003 IEEE International Workshop on*. IEEE, pp. 115–118.

[99] Rubins U, Grabovskis A, Grube J, Kukulis I (2008) Photoplethysmography analysis of artery properties in patients with cardiovascular diseases. In: *14th Nordic-Baltic Conference on Biomedical Engineering and Medical Physics*. Springer, pp. 319–322.

[100] Allen J, Murray A (2003) Age-related changes in the characteristics of the photoplethysmographic pulse shape at various body sites. *Physiological Measurement* 24: 297.

[101] Blaek R, Lee C (2010) Multi-resolution linear model comparison for detection of dicrotic notch and peak in blood volume pulse signals. *Analysis of Biomedical Signals and Images* 20: 378–386.

[102] Aiba Y, Ohshiba S, Horiguchi S, Morioka I, Miyashita K, et al. (1999) Peripheral hemodynamics evaluated by acceleration plethysmography in workers exposed to lead. *Industrial Health* 37: 3–8.

[103] Otsuka T, Kawada T, Katsumata M, Ibuki C (2006) Utility of second derivative of the finger photoplethysmogram for the estimation of the risk of coronary heart disease in the general population. *Circulation Journal* 70: 304–310.

[104] Zhang Y. et al. (2018) Evaluation of Cardiorespiratory Function During Cardiopulmonary Exercise Testing in Untreated Hypertensive and Healthy Subjects. *Frontiers in Physiology* 9: 1590.

[105] Sano Y (2003) Kasokudo myakuha ni kansuru kenkyuu no gaiyou. Technical report. 2004/02/23, in Japanese.

[106] Homma S, Ito S, Koto T, Ikegami H (1992) Relationship between accelerated plethysmogram, blood pressure and arteriolar elasticity. *Japanese Journal of Physical Fitness and Sports Medicine* 41: 98–107.

[107] Iokibe T, Kurihara M, Maniwa Y, Ohta S, Uchida I, et al. (2003) Chaos-based quantitative health evaluation and disease state estimation by acceleration plethysmogram. *Journal of Japan Society for Fuzzy Theory and Intelligent Informatics* 15: 565–576.

[108] Inoue N, et al. (2017) Second derivative of the finger photoplethysmogram and cardiovascular mortality in middle-aged and elderly Japanese women. *Hypertension Research* 40.2: 207–211.

[109] Elgendi M (2016) TERMA framework for biomedical signal analysis: An economic-inspired approach. *Biosensors* 6.4: 55.

[110] Pan J, Tompkins W (1985) A real-time QRS detection algorithm. *IEEE Transactions on Biomedical Engineering* 32: 230–236.

[111] Chen H, Chen S (2003) A moving average based filtering system with its application to real-time QRS detection. In: *Proceedings of the IEEE Computers in Cardiology.* pp. 585–588. doi:10.1109/CIC.2003.1291223.

[112] Chen SW, Chen HC, Chan HL (2006) A real-time QRS detection method based on moving-averaging incorporating with wavelet denoising. *Computer Methods and Programs in Biomedicine* 82: 187–195.

[113] Matsuyama A (2009) ECG and APG Signal Analysis during Exercise in a Hot Environment. PhD Thesis, Charles Darwin University, Darwin, Australia.

[114] Mattson CA, Mullur AA, Messac A (2004) Smart pareto filter: obtaining a minimal representation of multiobjective design space. *Engineering Optimization* 36: 721–740.

[115] Oppenheim A, Shafer R (1989) *Discrete-time Signal Processing.* Upper Saddle River, NJ: Prentice Hall.

[116] Firbank M, Coulthard A, Harrison R, Williams F. (1999) A comparison of two methods for measuring the signal to noise ratio on MR images. *Physics in Medicine and Biology* 44: 261–264.

[117] Li C, Zheng C, Tai C (1995) Detection of ECG characteristic points using wavelet transforms. *IEEE Transactions on Biomedical Engineering* 42: 21–28.

[118] Zong W, Heldt T, Moody GB, Mark RG (2003) An open-source algorithm to detect onset of arterial blood pressure pulses. In: *Proceedings of the IEEE Computers in Cardiology*. pp. 259–262.

[119] Elgendi M, Eskofier B, Dokos S, Abbott D (2014) Revisiting QRS detection methodologies for portable, wearable, battery-operated, and wireless ECG systems. *PLoS ONE* 9: e84018.

[120] Thakor NV, Webster JG, Tompkins WJ (1983) Optimal QRS detector. *Medical and Biological Engineering* 21: 343–350.

[121] Sahambi JS, Tandon S, Bhatt RKP (1997) Using wavelet transforms for ECG characterization. An on-line digital signal processing system. *IEEE Engineering in Medicine and Biology Magazine* 16: 77–83.

[122] Moraes JCTB, Freitas MM, Vilani FN, Costa EV (2002) A QRS complex detection algorithm using electrocardiogram leads. In: *Proceedings of the IEEE Computers in Cardiology*. pp. 205–208. doi:10.1109/CIC.2002.1166743.

[123] Mahmoodabadi SZ, Ahmadian A, Abolhasani MD (2005) ECG feature extraction using Daubechies wavelets. In: *Proceedings of the Fifth IASTED International Conference*. pp. 343–348.

[124] Holsinger W, Kempner K, Miller M (1971) QRS preprocessor based on digital differentiation. *IEEE Transactions on Biomedical Engineering* 18: 212–217.

[125] Okada M (1979) A digital filter for the QRS complex detection. *IEEE Transaction on Biomedical Engineering* 26: 700–703.

[126] Morizet-Mahoudeaux P, Moreau C, Moreau D, Quarante JJ (1981) Simple microprocessor-based system for on-line e.c.g. arrhythmia analysis. *Medical & Biological Engineering & Computing* 19: 497–500.

[127] Benitez DS, Gaydecki PA, Zaidi A, Fitzpatrick AP (2000) A new QRS detection algorithm based on the Hilbert transform. In: *Proceedigns of the IEEE Computers in Cardiology*. pp. 379–382.

[128] Arzeno N, Poon C, Deng Z (2006) Quantitative analysis of QRS detection algorithms based on the first derivative of the ECG. In: *Proceedings of the 28th Annual International Conference of the IEEE Engineering in Medicine and Biology Society*. pp. 1788–1791.

[129] Zhang F, Lian Y (2007) Electrocardiogram QRS detection using multiscale filtering based on mathematical morphology. In: *Proceedings of the 29th Annual International Conference of the IEEE Engineering in Medicine and Biology Society*. pp. 3196–3199.

[130] Arzeno N, Deng Z, Poon C (2008) Analysis of first-derivative based QRS detection algorithms. *IEEE Transactions on Biomedical Engineering* 55: 478–484.

[131] Friesen G, Jannett T, Jadallah M, Yates S, Quint S, et al. (1990) A comparison of the noise sensitivity of nine QRS detection algorithms. *IEEE Transactions on Biomedical Engineering* 37: 85–98.

[132] Englese WAH, Zeelenberg C (1979) A single scan algorithm for QRS detection and feature extraction. In: *Proceedings of the IEEE Computers in Cardiology*. pp. 37–42.

[133] Fraden J, Neuman M (1980) QRS wave detection. *Medical and Biological Engineering and Computing* 18: 125–132.

[134] Elgendi M, Howard N, Lovell N, Cichocki A, Brearley M, et al. (2016) A six-step framework on biomedical signal analysis for tackling noncommunicable diseases: current and future perspectives. *JMIR Biomedical Engineering* 1: e1.

[135] Christov II (2004) Real time electrocardiogram QRS detection using combined adaptive threshold. *Biomedical Engineering Online* 3: 28.

[136] Chiarugi F, Sakkalis V, Emmanouilidou D, Krontiris T, Varanini M, et al. (2007) Adaptive threshold qrs detector with best channel selection based on a noise rating system. In: *Proceedings of the IEEE Computers in Cardiology*, pp. 157–160. doi:10.1109/CIC.2007.4745445.

[137] Mann ME, Lees JM (1996) Robust estimation of background noise and signal detection in climatic time series. *Climatic Change* 33: 409–445.

[138] Chatterjee SK, Das S, Maharatna K, Masi E, Santopolo L, et al. (2015) Exploring strategies for classification of external stimuli using statistical features of the plant electrical response. *Journal of The Royal Society Interface* 12: 1–13.

[139] Koos C, Vorreau P, Vallaitis T, Dumon P, Bogaerts W, et al. (2009) All-optical high-speed signal processing with silicon–organic hybrid slot waveguides. *Nature Photonics* 3: 216–219.

[140] Furuya M, Hamano Y, Naito I (1996) Quasi-periodic wind signal as a possible excitation of chandler wobble. *Journal of Geophysical Research: Solid Earth* 101: 25537–25546.

[141] Stella L, Vietri M (1998) Lense-thirring precession and quasi-periodic oscillations in low-mass X-ray binaries. *The Astrophysical Journal Letters* 492: L59.

[142] Jerolmack DJ, Paola C (2010) Shredding of environmental signals by sediment transport. *Geophysical Research Letters* 37.

[143] Tandon N, Choudhury A (1999) A review of vibration and acoustic measurement methods for the detection of defects in rolling element bearings. *Tribology International* 32: 469–480.

[144] Alwan A, Armstrong T, Bettcher D, Branca F, Chisholm D, Ezzati M, et al. (2014) *Global Status Report on Noncommunicable Diseases 2014*. WHO/NMH/NVI/15.1. Geneva, Switzerland: World Health Organization.

[145] LeDuc P, Agaba M, Cheng CM, Gracio J, Guzman A, et al. (2014) Beyond disease, how biomedical engineering can improve global health. *Science Translational Medicine* 6: 266fs48.

[146] Dhawan AP, Heetderks WJ, Pavel M, Acharya S, Akay M, et al. (2015) Current and future challenges in point-of-care technologies: a paradigm-shift in affordable global healthcare with personalized and preventive medicine. *IEEE Journal of Translational Engineering in Health and Medicine* 3: 1–10.

[147] Jasemian Y, Arendt-Nielsen L (2005) Evaluation of a realtime, remote monitoring telemedicine system using the bluetooth protocol and a mobile phone network. *Journal of Telemedicine and Telecare* 11: 256–260.

[148] Jurik AD, Weaver AC (2008) Remote medical monitoring. *Computer* 41: 96–99.

[149] Baig M, Gholamhosseini H, Connolly M (2013) A comprehensive survey of wearable and wireless ECG monitoring systems for older adults. *Medical & Biological Engineering & Computing* 51: 485–495.

[150] Blahut RE (2010) *Fast Algorithms for Signal Processing*. Cambridge University Press.

[151] Orlowsky B, Seneviratne SI (2012) Global changes in extreme events: regional and seasonal dimension. *Climatic Change* 110: 669–696.

[152] Brearley M (2016) Cooling methods to prevent heat-related illness in the workplace. *Workplace Health & Safety* 64: 80.

[153] Lundgren K, Kuklane K, Gao C, Holmer I (2013) Effects of heat stress on working populations when facing climate change. *Industrial Health* 51: 3–15.

[154] Carter III R, Cheuvront SN, Williams JO, Kolka MA, Stephenson LA, et al. (2005) Epidemiology of hospitalizations and deaths from heat illness in soldiers. Technical report, DTIC Document.

[155] Xiang J, Hansen A, Pisaniello D, Bi P (2015) Extreme heat and occupational heat illnesses in south Australia, 2001–2010. *Occupational and Environmental Medicine.* 72: 580–586.

[156] Xiang J, Bi P, Pisaniello D, Hansen A (2014) The impact of heatwaves on workers health and safety in adelaide, south australia. *Environmental Research* 133: 90–95.

[157] Moran DS, Erlich T, Epstein Y (2007) The heat tolerance test: an efficient screening tool for evaluating susceptibility to heat. *Journal of Sport Rehabilitation* 16: 215–221.

[158] Binkley HM, Beckett J, Casa DJ, Kleiner DM, Plummer PE (2002) National athletic trainers' association position statement: exertional heat illnesses. *Journal of Athletic Training* 37: 329.

[159] Casa DJ, Becker SM, Ganio MS, Brown CM, Yeargin SW, et al. (2007) Validity of devices that assess body temperature during outdoor exercise in the heat. *Journal of Athletic Training* 42: 333.

[160] Bruce-Low SS, Cotterrell D, Jones GE (2006) Heart rate variability during high ambient heat exposure. *Aviation, Space, and Environmental Medicine* 77: 915–920.

[161] Yamamoto S, Iwamoto M, Inoue M, Harada N (2007) Evaluation of the effect of heat exposure on the autonomic nervous system by heart rate variability and urinary catecholamines. *Journal of Occupational Health* 49: 199–204.

[162] Galloway S, Maughan RJ (1997) Effects of ambient temperature on the capacity to perform prolonged cycle exercise in man. *Medicine & Science in Sports & Exercise* 29: 1240–1249.

[163] Crandall CG, Johnson JM, Convertino VA, Raven PB, Engelke KA (1994) Altered thermoregulatory responses after 15 days of head-down tilt. *Journal of Applied Physiology* 77: 1863–1867.

[164] Lisman P, Kazman JB, O'Connor FG, Heled Y, Deuster PA (2014) Heat tolerance testing: association between heat intolerance and anthropometric and fitness measurements. *Military Medicine* 179: 1339–1346.

[165] Tamura T, Maeda Y, Sekine M, Yoshida M (2014) Wearable photoplethysmographic sensors–past and present. *Electronics* 3: 282–302.

[166] Edwards J (2012) Taking the pulse of pulse oximetry in africa. *Canadian Medical Association Journal* 184: E244–E245.

[167] Kwok AC, Funk LM, Baltaga R, Lipsitz SR, Merry AF, et al. (2013) Implementation of the world health organization surgical safety checklist, including introduction of pulse oximetry, in a resource-limited setting. *Annals of Surgery* 257: 633–639.

[168] Lozano R, Naghavi M, Foreman K, Lim S, Shibuya K, et al. (2013) Global and regional mortality from 235 causes of death for 20 age groups in 1990 and 2010: a systematic analysis for the global burden of disease study 2010. *The Lancet* 380: 2095–2128.

[169] Hutcheon JA, Lisonkova S, Joseph K (2011) Epidemiology of pre-eclampsia and the other hypertensive disorders of pregnancy. *Best Practice & Research Clinical Obstetrics & Gynaecology* 25: 391–403.

[170] Wagner LK (2004) Diagnosis and management of preeclampsia. *American Family Physician* 70: 2317–2324.

[171] Payne BA, Hutcheon JA, Dunsmuir D, Cloete G, Dumont G, et al. (2015) Assessing the incremental value of blood oxygen saturation (spo 2) in the minipiers (pre-eclampsia integrated estimate of risk) risk prediction model. *Journal of Obstetrics and Gynaecology Canada* 37: 16–24.

[172] Kyriacou P, Moye A, Choi D, Langford R, Jones D (2001) Investigation of the human oesophagus as a new monitoring site for blood oxygen saturation. *Physiological Measurement* 22: 223.

[173] Webster JG (2003) *Electrical Measurement, Signal Processing, and Displays.* CRC Press.

[174] Shafique M, Kyriacou PA (2012) Photoplethysmographic signals and blood oxygen saturation values during artificial hypothermia in healthy volunteers. *Physiological Measurement* 33: 2065.

[175] Payne BA, Hutcheon JA, Ansermino JM, Hall DR, Bhutta ZA, et al. (2014) A risk prediction model for the assessment and triage of women with hypertensive disorders of pregnancy in low-resourced settings: the minipiers (pre-eclampsia integrated estimate of risk) multi-country prospective cohort study. *PLoS Medicine* 11: 1–13.

[176] Liang Y, Chen Z, Ward R, Elgendi M (2018) Hypertension assessment via ecg and ppg signals: an evaluation using mimic database. *Diagnostics* 8: 65.

[177] Martínez G, Howard N, Abbott D, Lim K, Ward R, et al. (2018) Can photoplethysmography replace arterial blood pressure in the assessment of blood pressure? *Journal of Clinical Medicine* 7: 316.

[178] Liang Y, Chen Z, Ward R, Elgendi M (2019) Hypertension assessment using photoplethysmography: a risk stratification approach. *Journal of Clinical Medicine* 8.

[179] Liang Y, Chen Z, Ward R, Elgendi M (2018) Photoplethysmography and deep learning: enhancing hypertension risk stratification. *Biosensors* 8.

[180] Elgendi M (2015) Pulse oximetry. Pre-empt Annual Report 2014–2015: 96.

[181] Clifford GD, Scott DJ, Villarroel M, et al. (2009) User guide and documentation for the mimic ii database. *MIMIC-II database version* 2.

Index

Page numbers in **bold** indicate tables, page numbers in *italic* indicate figures, page numbers followed by n indicate chapter notes, and page numbers followed by a indicate appendix.